THE STANDARD-VACUUM OIL COMPANY
AND UNITED STATES EAST
ASIAN POLICY, 1933–1941

In July 1940, American Ambassador to Japan Joseph C. Grew characterized United States economic policy in East Asia as "a sword of Damocles" over the head of Japan.* War had enveloped Europe, Japan had grown increasingly belligerent in East Asia, and the United States had abrogated its commercial treaty with Japan without clearly indicating whether or not it intended to invoke severe economic sanctions. Since Japan obtained four-fifths of its petroleum from the United States, its leaders had begun to contemplate the rich resources of the Netherlands East Indies as an alternative source of supply. A year after Grew penned his "sword of Damocles" comment Japanese troops occupied southern Indochina, the United States invoked a full trade embargo, and on December 7 Japan struck south toward the oil fields and east at the American fleet in Pearl Harbor. This study deals with events leading to that embargo from the viewpoint of an American company whose operations became deeply intertwined with the formulation and execution of American policy—the Standard-Vacuum Oil Company.

*Grew Diary, Vol. 101, p. 4438, Joseph C. Grew MSS, Houghton Library, Harvard University, Cambridge, Massachusetts.

IRVINE H. ANDERSON, JR.

THE STANDARD-VACUUM

OIL COMPANY AND

UNITED STATES

EAST ASIAN POLICY,

1933-1941

PRINCETON UNIVERSITY PRESS

To Emily and Grace

CONTENTS

ACKNOWLEDGMENTS

So MANY people have helped in so many ways to make this study possible that it would be difficult to acknowledge my full indebtedness to all of them. I am especially grateful to Dr. Daniel R. Beaver for his encouragement all along the way and for his initial suggestion that this was a topic worth pursuing. Dr. Henrietta M. Larson was most helpful as I groped my way through the early stages of research, and none of the material from the Exxon Corporation would have been available without the gracious assistance of Mr. Charles E. Springhorn.

Much of the research for this study was made possible by a Woodrow Wilson Dissertation Fellowship, and by a generous travel grant from the University of Cincinnati. The courtesy of archivists in many locations made the work much more pleasant, and I am especially indebted to the staffs of the National Archives in Washington and the Public Record Office in London for their patience and helpfulness. I would also like to thank the Yale University Library for permission to quote from the papers of Henry L. Stimson, and the Harvard College Library for permission to quote from the papers of Joseph C. Grew, W. Cameron Forbes, and Jay Pierrepont Moffat.

I am deeply grateful to a number of people who read the original manuscript and offered excellent suggestions for its improvement, each from a different viewpoint: Dr. George E. Engberg, Dr. Akira Iriye, Dr. Han-kyo Kim, Dr. Barbara N. Ramusack, and especially Dr. Mira Wilkins. Obviously, however, any remaining errors of fact or interpretation are my responsibility alone.

Last, but by no means least, I would like to acknowledge my deepest debt of all, to my daughters Emily and Grace, who will understand why this book is dedicated to them.

LIST OF TABLES

LIST OF TABLES

THE STANDARD-VACUUM OIL COMPANY
AND UNITED STATES EAST
ASIAN POLICY, 1933–1941

INTRODUCTION

In the decade before Pearl Harbor, the Standard-Vacuum Oil Company occupied a position unique in the Asian oil trade and in that portion of Asia threatened by Japanese expansion.[1] With marketing outlets in both Japan and China and supplies drawn largely from the Netherlands East Indies, "Stanvac" had good reason to be apprehensive of Japanese intentions. Its position in the Asian oil trade also enabled the company to involve itself effectively in the regulation of Japan's oil supply whenever a policy could be agreed upon with Royal Dutch-Shell and the American government. Since the immediate causes of World War II in the Pacific included a *de facto* American embargo on oil to Japan which started in July 1941, and a Japanese resolve to seize the necessary supplies in the Indies, the company's operations inevitably involved it in events leading to the confrontation between Japan and the United States. In this context the question logically arises as to what role, if any, Stanvac itself played in the formulation and execution of American East Asian policy before Pearl Harbor. This study is an attempt to deal with that question.

Formed in 1933 as a consolidation of the Asian producing and refining operations of the Standard Oil Company (New Jersey), later Exxon, with the Asian distribution network of Socony-Vacuum, later Mobil, the Standard-Vacuum Oil Company functioned as a relatively independent joint subsidiary of those two corporations until its dissolution for business and legal reasons in 1962.[2] Its market included South and East Africa, India, Australia, Southeast Asia, China, and Japan, with roughly half its supplies drawn from production and re-

[1] This summary of Stanvac's position in Asian oil is based on data incorporated in Appendices A and B.

[2] The name "Stanvac" survived past 1962. "P. T. Stanvac Indonesia" continued as a jointly owned production subsidiary after the rest of the original Stanvac assets had been divided between the two parent corporations.

3

fining operations in the Netherlands East Indies and the balance purchased chiefly from other affiliates of the two parent corporations. Prior to World War II Stanvac was the only American oil company operating a distribution business within both Japan and China and the only one with a major supply source in the Netherlands Indies. Caltex (owned jointly by Standard of California and The Texas Company) maintained a slightly smaller distribution business in China but none in Japan, and had begun exploration in the Indies but had not yet commenced production there at the outbreak of war. A number of American oil companies sold to Japan, but their business was almost entirely in bulk shipments from the United States to Japanese importers and distributors. Only the British- and Dutch-owned Royal Dutch-Shell occupied a position roughly parallel to that of Stanvac in East and Southeast Asia, with production in the Indies and major distribution outlets in both Japan and China. Shell's operation in all three areas was larger than that of Stanvac, but Stanvac's godowns (warehouses), storage tanks, deepwater terminals, wells, and refineries combined, constituted the single largest American direct investment in East and Southeast Asia prior to Pearl Harbor.

In strategic terms, Stanvac's position was also unique. Japan's own petroleum resources were woefully inadequate to her military and industrial needs, and during the 1930s an increasing portion of her requirements were purchased from American sources. By 1939 eighty percent of Japan's imported petroleum and petroleum products came from the United States, with most of the rest procured from the Netherlands Indies, where production was almost completely controlled by Shell and Stanvac. As if to bait the trap, sufficient petroleum to make Japan independent of American supplies would be available if Japan ever gained complete control of production in the Indies. Thus any move to cut off her supply of American oil was likely to trigger a Japanese thrust toward the Indies. Stanvac itself was not the major supplier of oil to Japan, but in this delicate situation it was in a position to cooperate effectively with Shell

4

and the American government to maintain or curtail the flow of oil to Japan whenever the three decided to act in unison. The surviving records clearly demonstrate that considerable teamwork did, in fact, develop in the late 1930s between Stanvac, the American State Department, Royal Dutch-Shell, and the British Foreign Office, but the question of who actually influenced whom and why requires more thorough analysis.

The reader should be aware of several constraints that surround this study and influence the shape it has taken. First is the problem of primary sources. No internal records of the Standard-Vacuum Oil Company itself could be located, although a few related documents were found and made available to the author by the Exxon Corporation. Stanvac's original records (except for a few legal and financial papers), are understood to have been destroyed shortly after it was reorganized out of existence in 1962. Fortunately there are extensive records of discussions and correspondence with Stanvac officers in the files of the Department of State deposited in the National Archives, and similar documents on contacts with Royal Dutch-Shell are in the records of the British Foreign Office at the Public Record Office in London. These two sources provide the foundation for this study, with extensive supplementary material from a wide variety of other locations. The study, therefore, contains considerably more detailed information on the interaction between Stanvac and governmental agencies than on Stanvac's internal policy-making process.

The organization of the narrative itself reflects two parallel conceptual frameworks. The Stanvac story is a fascinating one in its own right, but it also illustrates certain organizational trends in modern industrial societies and is intricately interwoven with the origins of World War II in the Pacific. Neither of these conceptual frameworks is dealt with directly in the narrative, but both have significantly influenced the structure of the study. Since there is undeniable validity in the notion that "truth is in the eye of the beholder," it seems only fair to share

5

with the reader these two perceptual filters through which this story has passed.

The study is based on a conviction that many institutions in modern industrial societies have steadily evolved toward the rational, impersonal, "bureaucratic" structure described in theoretical terms by Max Weber.[3] Within the Weberian framework, however, there are paradoxes in organizational behavior which have been well described by later theorists, especially the French sociologist Michel Crozier.[4] Crozier agrees that the trend toward formal structure has been unrelenting as a product of human attempts to reduce uncertainty and increase predictability in organizational behavior. But he asserts that behavior within organizations cannot be explained by reference to formal structure alone. Crozier emphasizes power relationships, the informal organization, and the role of managers as political leaders within organizations rather than emphasizing their roles as true entrepreneurs or technicians. In short, he confirms the trend toward increasingly complex organizations, but suggests that their actual functioning must be understood in terms of both formal and informal power relationships.[5] In modern organizations, Crozier argues, "the power of A over B

[3]Max Weber, *The Theory of Social and Economic Organization*, translated by A. M. Henderson and Talcott Parsons (New York: The Free Press, 1964), pp. 329–41; and *From Max Weber: Essays in Sociology*, translated, edited, and with an introduction by H. H. Gerth and C. Wright Mills (New York: Oxford University Press, 1946), pp. 196–244. For criticism of Weber's somewhat static model, see especially Peter M. Blau, *The Dynamics of Bureaucracy* (Chicago: University of Chicago Press, 1955) and Amitai Etzioni, *A Comparative Analysis of Complex Organizations* (New York: The Free Press, 1961).

[4]Michel Crozier, *The Bureaucratic Phenomenon*, translated by the author (Chicago: University of Chicago Press, 1967), pp. 145–208, 296–300. In addition to original research and formulation of theory, Crozier has synthesized a considerable body of work by others in the field.

[5]Ibid., pp. 156–57. Crozier used Robert Dahl's definition of power as the ability of one person to cause another to do something he would not otherwise have done; see Robert A. Dahl, "The Concept of Power," *Behavioral Science*, II (July 1957), 201–15. Dahl has presented a

depends on A's ability to predict B's behavior and on the uncertainty of B about A's behavior. As long as the requirements of action create situations of uncertainty, the individuals who face them have power over those who are affected by the results of their choice."[6] While not overlooking other stimuli, he argues that much organizational behavior is explained by attempts to maintain or broaden one's own area of discretion and reduce the amount of uncertainty in what others may do. This theoretical viewpoint provides a useful basis for analysis of complex organizational situations and actions.

Crozier's observations appear especially useful to an understanding of the development of the petroleum industry in general and Stanvac in particular.[7] "Rationalization" in the petroleum industry helped create the fully integrated oil com-

---

refinement of essentially the same concept in his article, "Power," in David L. Sills, ed., *International Encyclopedia of the Social Sciences* (17 vols.; New York: Macmillan, 1968), XII, 405–14; and in his introductory survey, *Modern Political Analysis* (2nd ed.; Englewood Cliffs: Prentice-Hall, 1970), pp. 14–34.

[6]Crozier, p. 158.

[7]On the petroleum industry in general, useful background is contained in Harold F. Williamson *et al., The American Petroleum Industry* (2 vols.; Evanston: Northwestern University Press, 1959–63), Volume I, Harold F. Williamson and Arnold R. Daum, *The Age of Illumination 1859–1899,* and Volume II, Harold F. Williamson, Ralph L. Andreano, Arnold R. Daum, and Gilbert C. Klose, *The Age of Energy 1899–1959;* John G. McLean and Robert W. Haigh, *The Growth of Integrated Oil Companies* (Boston: Graduate School of Business Administration, Harvard University, 1954); Gerald D. Nash, *United States Oil Policy, 1890–1964: Business and Government in Twentieth Century America* (Pittsburgh: University of Pittsburgh Press, 1968); and in the critical analysis by Edith T. Penrose, *The Large International Firm in Developing Countries: The International Petroleum Industry* (London: Allen and Unwin, 1968). The three volume *History of Standard Oil Company (New Jersey)* contains a wealth of relevant information, especially the recently published third volume by Henrietta M. Larson, Evelyn H. Knowlton, and Charles S. Popple, *New Horizons 1927–1950* (New York: Harper & Row, 1971). The two earlier volumes are Ralph W. Hidy and Muriel E. Hidy, *Pioneering in Big Business 1882–1911* (New York: Harper, 1955), and George S. Gibb and Evelyn H. Knowlton, *The Resurgent Years, 1911–1927* (New York: Harper & Brothers, 1956).

7

pany, combining control of production, refining, transportation, and marketing as a means for achieving maximum efficiency and profitability. And an urge toward rationalization provided much of the impetus behind recurring attempts to reach formal and informal cooperative agreements between oil companies on both a national and an international level. If nothing else, predictable markets could ensure optimum utilization of expensive facilities. All of the data in Chapter I, describing the origins of Stanvac and Royal Dutch-Shell, have been organized around this theme.

Paradoxically, none of the managers involved behaved like technicians or efficiency experts. All were colorful, resourceful, and eager to respond to unforeseen problems. Stanvac board chairman George Walden always had at his fingertips a wealth of statistics which could only have been produced by a highly efficient formal organization, but he relied heavily on personal contacts for results and appeared to enjoy the game immensely. Walden's chief contact in the State Department, Far Eastern Division chief Stanley Hornbeck, was an expert at delineating formal areas of responsibility, but he carried on a lively correspondence outside of formal channels with scores of individuals in business and government. The high-flown language of the resulting voluminous correspondence sometimes made it difficult to follow the actual trend of events, and Crozier's theory on bureaucratic power was used as an analytical tool to cut through the maze and assess the more critical power relationships. Specifically, the question was posed as to which party had the most influential administrative control over concrete actions affecting the flow of British and American oil to Japan in each of the following relationships: the position of the British government vis-à-vis the government of the United States throughout the decade, the position of Stanvac vis-à-vis the American government in 1940 and 1941, and the position of the State Department vis-à-vis the Treasury Department in 1940 (before the freeze of Japanese assets in the United States) and again in 1941 (after the financial freeze). In each case, it became clear that the views that prevailed were those of the party with the most freedom to make key administrative decisions—just as Crozier had theorized. Most obvious was the fact that the British government consistently had to defer to the American government for the simple reason that most of Japan's oil came from the United States and not Britain. Some of the other relationships were more intricate and become apparent only as the narrative is traced in detail, but the results of this analytical approach directly influenced the interpretation offered in Chapters V and VI.

Finally, there is the question of where this study fits in the broader history of the period. The Stanvac episode is only one segment of the whole story of the coming of World War II in the Pacific, but it illustrates effectively several facets of that tragic sequence of events. The author's views on the origins of the Pacific war are derived from original research, but rely heavily on the work of Iriye, Crowley, Butow, Lu, Jones, Borg, Langer and Gleason, Feis, and Schroeder.[8] In broad concept,

[8]Akira Iriye, *After Imperialism: The Search for a New Order in the Far East, 1921–1931* (Cambridge: Harvard University Press, 1965); James B. Crowley, *Japan's Quest for Autonomy: National Security and Foreign Policy, 1930–1938* (Princeton: Princeton University Press, 1966); Robert J. C. Butow, *Tojo and the Coming of War* (Princeton: Princeton University Press, 1961); David J. Lu, *From the Marco Polo Bridge to Pearl Harbor: Japan's Entry Into World War II* (Washington, D.C.: Public Affairs Press, 1961); Francis C. Jones, *Japan's New Order in East Asia: Its Rise and Fall, 1937–1945* (London: Oxford University Press, 1954); Dorothy Borg, *The United States and the Far Eastern Crisis of 1933–1938: From the Manchurian Incident Through the Initial Stage of the Undeclared Sino-Japanese War* (Cambridge: Harvard University Press, 1964); William L. Langer and S. Everett Gleason, *The Challenge to Isolation, 1937–1940* (New York: Harper & Brothers, 1952), and *The Undeclared War, 1940–1941* (New York: Harper & Brothers, 1953); Herbert Feis, *The Road to Pearl Harbor: The Coming of the War Between the United States and Japan* (Princeton: Princeton University Press, 1950); and Paul W. Schroeder, *The Axis Alliance and Japanese-American Relations, 1941* (Ithaca: Cornell University Press, 1958). See also the excellent collection of essays in Dorothy Borg and Shumpei Okamoto, eds., *Pearl Harbor as History: Japanese-American Relations, 1931–1941* (New York: Columbia University Press, 1973).

they most nearly coincide with the interpretation advanced by Akira Iriye in his extended essay, *Across the Pacific.*[9]

Setting aside scores of contributing factors, the Pacific war appears to have originated in a combination of conflicting ideologies, historical accident, and human miscalculation. At least since the 1890s Japan and the United States had both been on an expansionist course in East Asia, and in retrospect it is hardly surprising that the two eventually came into conflict. Within this framework, however, there were more specific irritants. For one thing, the Japanese theory of how East Asia should be organized politically and economically was diametrically opposed to the thinking of most Americans. The Japanese view, widely held by civilian and military officials alike, perceived Japan as something of a Confucian elder brother to the rest of Asia and linked this with what James Crowley has termed a Japanese "quest for autonomy" in a hostile world. This set of ideas had a surprising degree of internal consistency, and found expression in the notion of a "Greater East Asia Co-Prosperity Sphere" which evolved in the late 1930s. Alone, this was not a cause for war, but it conflicted with the deep-rooted American "Open Door Policy" in China. By the 1930s most American diplomats and business men involved with East Asia were wedded to the idea that the affairs of foreign powers in China should be governed by strict legalistic rules of international conduct which guaranteed Chinese territorial integrity and equality of commercial opportunity therein. The two concepts involved differing philosophical traditions—the Confucian "rule of man" versus the Western "rule of law"—but they were inseparable from what best served the economic interests of each side. Idealism and self-interest form a potent combination in any culture. In practical terms, these two concepts resulted in increasing friction between Japan and the United States over a host of incidents arising from Japanese expansion into China.

[9] Akira Iriye, *Across the Pacific: An Inner History of American-East Asian Relations* (New York: Harcourt, Brace & World, 1967), pp. 200–226.

By coincidence this friction developed at a point in the history of technology where petroleum had begun to assume a major role in the fueling of naval vessels, tanks, and aircraft. While the navies of the world had begun conversion from coal to oil shortly before World War I, petroleum really came into its own as a military fuel in the interwar period. As already noted, this development found Japan critically short, dependent on the United States for much of its supply, and lured by resources ample to its needs in Southeast Asia. Emerging technology thus added an irritant but still did not provide a cause for war.

To this may be added the historical coincidence of the war in Europe. With the fall of France to Hitler in June of 1940, the likelihood of Japanese-American conflict mounted significantly. The United States began to perceive its security as inseparable from that of Britain. While Europe was given first priority in military planning from late 1940, the Americans also became concerned about the Commonwealth base at Singapore and the tin, rubber, and oil of Southeast Asia. From the Japanese viewpoint, the fall of France and the Netherlands together with the preoccupation of Britain laid bare the Western colonies in Southeast Asia and created a never to be repeated opportunity for Japan to realize its desire for economic and strategic autonomy. At the same time that the United States had begun to link its own security to that of Southeast Asia, Japan was sorely tempted to strike toward that area. The United States had terminated its commercial treaty with Japan on January 26, 1940, without clearly indicating whether or not it intended to invoke severe economic sanctions, and this "sword of Damocles" did nothing to allay Japanese concern.

At this point a type of circular reasoning gripped Japanese leaders, especially in the previously cautious Navy. "The policy of southern advance would make conflict with the United States inevitable; since war was bound to occur, Japan should advance southward to prepare for the conflict."[10] The leaders

[10] Ibid., p. 208.

11

never appeared to have recognized that it was their own actions that were creating hostility in the United States, and were genuinely surprised when their occupation of southern Indochina in July 1941 triggered a complete American embargo on trade—including oil. On the American side, the most widely held school of thought argued that Japan could be checked by firm talk and economic sanctions, but opponents of an embargo held the actual reins up to July of 1941. The move into Indochina appears to have convinced them that Japan would continue southward under any circumstances and that it would be foolish to continue fueling her war machine. The advocates of an embargo therefore prevailed, but neither group appears to have realized that this forced Japan's back to the wall as her supply of oil dwindled, and she finally struck out with a sense of desperation. In the final analysis, both sides reacted to what they perceived as threats to their own security, and neither fully understood the other's viewpoint. Whether different perceptions would have changed the outcome is problematic; history cannot be replayed.

The Standard-Vacuum Oil Company was involved to a greater or lesser degree in every phase of the process outlined. At issue is the question of whether its role was one of observer, implementor, or instigator of decisions. Stanvac's relationship to this process was a complex one—especially in 1941—and an attempt has been made to provide sufficient data for the reader to draw his own conclusions, even if they differ from those of the author. To the extent that generalizations are possible, however, this study suggests that during the 1930s Stanvac's exposed position was to some slight degree responsible for the deterioration in relations. It sought and received American diplomatic support for protection of its marketing outlets in Japan and China, while Japan attempted to achieve autonomy in petroleum refining and distribution. Although Stanvac itself simply wished to protect a legitimate business investment, the series of events increased American diplomatic irritation over Japanese violations of the Open Door and served as one more indicator to Japan that she was seeking autonomy in an essentially hostile world. Beginning in 1938, however, Stanvac management became convinced that the more serious threat lay in a possible Japanese attack on the Netherlands East Indies, and the company shifted its emphasis to full support of American policy in Southeast Asia—partially to gain protection for its heavy investment in that area. During 1940 and early 1941 the company played a key role in implementation of an American policy to avoid—or at least postpone—a confrontation by permitting Japan a carefully controlled flow of oil from both the Indies and the United States. But in July 1941, when Japanese occupation of southern Indochina convinced many Americans that further temporizing was pointless, the policy of restraint and the tightly knit team of which Stanvac had become a part disintegrated simultaneously. During the month of confusion that followed the freeze of Japanese funds, Stanvac participated along with a number of governmental agencies in a bureaucratic tangle which slowly congealed into an unplanned *de facto* embargo—cutting Japan off from its supply of foreign oil and starting the clock ticking on the final decision for war. In short, this study concludes that Stanvac's role shifted from instigator to implementor of American oil policy as events moved toward a final confrontation.

ONE

THRUST, PARRY, LUNGE, AND
COMPROMISE

NO BETTER term describes the evolution of Standard Oil's position in Asia prior to 1933 than "metamorphosis"—the gradual transformation of its role into something totally different from its original form. When Standard's predecessors first entered the China trade in the 1860s they did so as merchants of kerosene, a product whose chief value was household illumination. Step by step over the next seventy years, as a result of competitive pressure and the emergence of petroleum as a military fuel, Standard Oil's role was transformed into something quite different. By 1933 the newly formed Standard-Vacuum Oil Company shared with Royal Dutch-Shell virtually complete control of the most promising source of petroleum in Asia and together with the American government was in a position to regulate supplies to an increasingly belligerent Japan. Such a development could hardly have been foreseen in the days of the kerosene trade, but the story of how this transformation came about provides considerable insight into developments subsequent to 1933. More specifically, it illuminates the process by which Stanvac and Shell acquired so strategic a position, and illustrates the strong precedent for cooperation, which had been established long before Japanese pressure made collaboration imperative.

IF ONE symbol could be chosen to represent the early American oil trade in the Orient, it would undoubtedly be the "Mei Foo" kerosene lamp. Long before oil from the Netherlands Indies became significant in the Asian trade, American kerosene sales to China faced competition from other kerosene merchants, and the Mei Foo lamp represented an ingenious Standard Oil attempt to capture that market at the grass roots level.[1] Simple, efficient, and inexpensive, the lamp was intended to make it easy for Chinese peasants to switch from vegetable oil to kerosene for lighting and was either sold or given away by merchants throughout China around the turn of the century. "Mei Foo," the company name adopted by Standard Oil in China, translated roughly as "beautiful confidence," so the lamp served the triple purpose of creating potential customers, popularizing the company name, and advertising the virtues of the product.[2] This resourceful and aggressive adaptation to local conditions was typical of the American approach to the Asian oil trade from its beginnings in the early 1860s.

The story actually begins in 1859, for the American petroleum industry dates its origin from the discovery in that year of Drake's well at Titusville, Pennsylvania. Exports played an especially important role in petroleum's first half-century. In 1866 sixty-nine percent of total American refinery output was exported, and foreign sales of kerosene still accounted for fifty-eight percent of American output of that product in 1899.[3] Most of the trade passed through New York, but for years petroleum exports were handled almost as domestic transactions through commission merchants and agents of foreign importers, with control ending at water's

[1]Hidy and Hidy, pp. 259–62. A concise account of Standard's entrance into the Asian oil business is contained in Mira Wilkins, *The Emergence of Multinational Enterprise: American Business Abroad from the Colonial Period to 1914* (Cambridge: Harvard University Press, 1970), pp. 62–64, 82–87. For perspective on how this related to the general movement of Western business into the Orient, see two studies by George C. Allen and Audrey G. Donnithorne, *Western Enterprise in Far Eastern Economic Development: China and Japan* (London: Allen and Unwin, 1954), pp. 99–101, 205–206, and *Western Enterprise in Indonesia and Malaya: A study in Economic Development* (London: Allen and Unwin, 1954), pp. 174–80.

[2]"Half A Century In China," *Socony-Vacuum News*, February 1941, pp. 5–7; Mobil Oil Corporation Library, New York, hereafter cited as "Mobil Library."

[3]Williamson *et al.*, I, 322, 633.

edge.[4] Wooden barrels served as standard shipping containers for crude, and two five-gallon metal cans packed in a wooden case proved highly popular for kerosene, especially when shipped to areas with limited transport facilities, such as China.[5] The cases could be carried anywhere, and buyers found scores of uses for the metal cans.

Standard's interest in the export trade dated almost from John D. Rockefeller's entrance into the oil business. Within one year of the acquisition of his first refinery in Cleveland in 1865, Rockefeller and his younger brother William organized a company with offices in New York City specifically to handle export sales.[6] When the Standard Oil Trust was formed in 1882 as a holding instrument for some forty companies, William's business was incorporated as the Standard Oil Company of New York, and remained the conduit for most foreign sales.[7] For a number of years the Trust coordinated the business of its multiple enterprises through a committee system, and foreign sales policy was set by an export committee meeting regularly in New York.[8] Since American oil dominated the world market in the early years and since Standard handled ninety percent of American exports,[9] this committee proved to be a highly influential body. European sales absorbed most of its attention, but Asia was not overlooked. As early as 1882 William H. Libby went abroad to study Oriental markets and promote the sales of Standard products.[10] Along with other efforts, Libby wrote a long pamphlet for translation into Chinese extolling

[4]Ibid., I, 326, 331; Hidy and Hidy, p. 124.

[5]Williamson *et al.*, I, 330; Hidy and Hidy, p. 124.

[6]Allan Nevins, *Study in Power: John D. Rockefeller, Industrialist and Philanthropist* (2 vols; New York: Scribner's, 1953), I, 37–40.

[7]Hidy and Hidy, pp. 46, 49, 54, 127.

[8]Ibid., pp. 125–28. The export committee actually controlled only crude, kerosene, and naptha, but these comprised the bulk of Standard's exports. Lubricating oil and paraffin wax came under jurisdiction of a separate committee which provided only loose coordination of those sales.

[9]Hidy and Hidy, p. 129; Nevins, II, 114–15.

[10]Hidy and Hidy, p. 137; Nevins, II, 114–15.

the virtues of kerosene for illumination. The pamphlet itself has not survived, but Libby's line of reasoning is preserved in an excerpt of a letter to the governor general of India:

I may claim for petroleum that it is something of a civilizer, as promoting among the poorest classes of these countries a host of evening occupations, industrial, educational, and recreative, not feasible prior to its introduction; and if it has brought a fair reward to the capital ventured in its development, it has also carried more cheap comfort into more poor homes than almost any discovery of modern times.[11]

This benign scene was marred in the mid-1880s by the onset of what came to be called the great Russian "oil war." Serious exploitation of petroleum deposits near Baku in southern Russia began about fifteen years after the opening of the American Pennsylvania fields, but the real impact of Russian oil was not felt until completion of the Trans-Caucasian rail line from Baku to Batoum in 1883.[12] This provided an outlet through a deep-water Black Sea port, permitting Russian petroleum products to challenge Standard by rapidly acquiring a sizable share of the European market. The high flow rate of Baku wells cut the cost of Russian crude, transportation cost less since the market was closer, and the Swedish Nobel brothers (Alfred, Robert, and Ludwig) introduced technical improvements such as continuous distillation into the Russian field quite early.[13] In fact, it was the enterprise of the Nobels coupled with the financial resources of the French Rothschilds that really launched Russian oil onto the world scene.[14] The drive proved so successful that, starting from practically zero

[11]Letter, William H. Libby to His Excellency the Viceroy and Governor-General of India, December 21, 1882, reprinted in U.S. Department of State, *Consular Reports,* "American Petroleum in Foreign Countries," No. 37 (January 1884), p. 406.

[12]Williamson *et al.,* I, 509, 630, 635.

[13]Ibid., I, 509–19, 630–35; Hidy and Hidy, pp. 131–36.

[14]Ibid., pp. 133–34; Wilkins (1970), p. 63; Nevins, II, 119–26.

in 1883, Russian kerosene captured twenty-two percent of the world market by 1889 and reached as high as thirty-four percent in 1899.[15] Although concentrated in Europe, the impact of the Nobels and Rothschilds touched Asia as well. Standard's export committee became increasingly concerned.

American interests could have responded by negotiating some general agreement on markets and prices with the Nobel-Rothschild interests, and Standard did, in fact, engage in inconclusive discussions of this type.[16] For its major effort, however, Standard chose to counterattack by meeting competition with more competition. To offset a similar move in England by the Nobel-Rothschild group in 1888, Standard organized the Anglo-American Oil Company, Ltd., to set up its own direct distribution network in Britain.[17] Anglo-American acquired a fairly broad charter, becoming in 1890 the instrument for a major policy change in the Orient. Largely at the instigation of Libby, Anglo-American began in that year to consign products in its own name to commission merchants east of Suez.[18] In effect, this ended the practice of f.o.b. sales to export merchants in American ports and extended control over

[15]Williamson *et al.*, I, 633. American petroleum accounted for almost all of the remainder, with Standard handling ninety percent of that. Comparable statistics are not available for products other than kerosene, but the literature suggests an equivalent growth in the Russian share of the market for other products as well. Russian kerosene sales in the Orient were originally handled chiefly by the British firm of Lane and McAndrew (Williamson *et al.*, I, 663).

[16]The extent to which agreement was actually reached remains in doubt. Nevins (II, 123–24) records a letter from John D. Archbold to Rockefeller in 1892 reporting a "tentative understanding" with the Rothschilds, but concludes that "in the end it proved impossible to reach any agreement." Hidy and Hidy (p. 237) conclude that "no general agreement was ever reached," although several Standard affiliates did reach "understandings covering individual markets." Williamson (I, 650) records numerous rumors of a pact in the 1890s but agrees that "no effective agreement providing for a global division of marketing territories was reached." See also Wilkins (1970), p. 82.

[17]Hidy and Hidy, pp. 147–48.

[18]Ibid., p. 152.

shipments until they reached Singapore, Shanghai, Yokohama, and other eastern ports. From there it was only a short step to development of a distribution network throughout East Asia, a task assigned to Standard Oil of New York in 1893.[19] New York Standard began an intensive program of building storage depots, appointing local representatives, and adjusting prices to undersell the competition. The Mei Foo lamp thus constituted one small phase of a new, aggressive marketing policy, and heralded the beginning of what became a well-developed distribution network and part of the largest single American direct investment in East and Southeast Asia in the 1930s.[20]

STANDARD had scarcely begun to adjust to competition from Russian oil when a flank attack developed under the leadership of two talented and resourceful organizers—Marcus Samuel and Henri Deterding. Within a relatively short period Samuel moved from being a general merchandise broker in the Anglo-Japanese trade, to dealer in Russian kerosene, to founder of the Shell half of what became Royal Dutch-Shell in 1907. Deterding began as the Penang subagent for a Netherlands trading firm, joined a struggling Indies production and refining company to help increase sales, and moved to leadership of the Royal Dutch half of Royal Dutch-Shell. The alliance thus created came to dominate the Asian oil trade and sparked Standard's interest in acquiring oil fields in the Indies in order to remain competitive in the Orient.

In 1874 Marcus Samuel and his brother had inherited a well-established brokerage firm that concentrated on the sale of manufactured items in the East and the purchase of jute, tea, and rice for sale in Europe.[21] By the late 1880s Japan had

[19]Ibid., pp. 261, 547–48.

[20]For a description of what became the East Asian distribution network of the Standard-Vacuum Oil Company, see Appendix A.

[21]Kendall Beaton, *Enterprise in Oil: A History of Shell in the United States* (New York: Appleton-Century-Crofts, 1957), pp. 38–39. A concise account of the origins of the Shell Transport and Trading Company is also given in Williamson *et al.*, I, 664–74.

launched an intense program of modernization under the Meiji Restoration, and the Samuel brothers began to search for products to replace the manufactured items that Japan would soon be able to provide for herself. Marcus became intrigued with the possibilities of kerosene, since that was not produced in Japan and was proving increasingly popular with Japanese consumers as a source of illumination.[22] In 1890 he traveled to Batoum for a conference with managers of the Rothschild's Société Commerciale et Industrielle de Naphte Caspienne et de la Mer Noire (usually referred to as "Bnito") and proposed the establishment of a bulk distribution system for Russian kerosene in the Orient. Bnito agreed to supply the kerosene f.o.b. Batoum, and Samuel arranged for construction of a tanker fleet, talked the Suez Canal Authority into changing its rules to permit the passage of tankers, and organized a Tank Syndicate of merchants in all major Asian ports to build storage terminals at the other end.[23] In July 1892, the first tanker passed through the Suez Canal eastbound with a cargo of kerosene, and a new marketing organization was born. Five years later the Tank Syndicate reorganized itself into the Shell Transport and Trading Company, incorporating the petroleum handling facilities of nine widely-dispersed trading companies.[24]

Parallel with this development a production and refining organization had begun to emerge in the Netherlands East Indies. Evidence of petroleum deposits had existed for years, and prospecting began shortly after the American boom sparked interest, but the first test drilling did not commence until 1871, and significant commercial production did not be-

[22]Beaton, p. 39; Williamson *et al.,* I, 665.

[23]Beaton, pp. 39–42; Williamson *et al.,* I, 665–68.

[24]Beaton, p. 43; Williamson *et al.,* I, 674. Of the nine merchant firms participating in the formation of Shell Transport and Trading, two were in London and one each in Germany, India, the Straits Settlement, French Indochina, Hong Kong, the Philippines, and Japan. The name "Shell" was taken from the brand name of the Tank Syndicate's kerosene, which in turn had been inspired by the fact that one of Samuel and Company's original trading items had been sea shells.

gin until 1894.[25] Numerous individuals participated in early pioneering ventures, but what proved to be the most successful originated in 1880 when Aeilko J. Zijlker of the East Sumatra Tobacco Company decided to investigate oil seepage in northern Sumatra. He found it to be of high quality, obtained permission from the Sultan of Langkat to work the area, raised capital, and enlisted the aid of the Dutch government to drill his first test well.[26] Although the results of several years' work proved inconclusive, Zijlker became convinced of the commercial possibilities of Sumatran oil and left the Indies for the Netherlands to raise funds. His trip sparked formation in 1890 of the Naamlooze Vennootschap Koninklijke Nederlandsche Maatschappij tot Exploitatie van Petroleumbronnen in Nederlandsche-Indië (otherwise known as the Royal Dutch Company), which bought out his concession, purchased the necessary equipment, and sent J. B. August Kessler to Sumatra to commence operations.[27] Kessler succeeded in getting several wells drilled, built a refinery, and became managing director of the company in 1892.[28] Four years later he persuaded a friend to leave a position as Penang subagent for the Netherlands Trading Company and take charge of sales for Royal Dutch. That friend was Henri Deterding, who assumed the position of managing director on Kessler's death in 1900.[29] Deterding thus came to head a promising production operation while Marcus Samuel commanded a far-flung marketing organization, and both feared competition from Standard Oil. In retrospect an alliance between the two appears quite logical,

[25]Alex L. Ter Braake, *Mining in the Netherlands East Indies* (New York: Institute of Pacific Relations, 1944), pp. 66–67; James W. Gould, *Americans in Sumatra* (The Hague: Martinus Nijhoff, 1961), p. 45; Allen and Donnithorne, *Western Enterprise in Indonesia and Malaya*, pp. 174–84; Williamson *et al.*, I, 631.

[26]F. C. Gerretson, *History of the Royal Dutch* (4 vols.; Leiden: E. J. Brill, 1953–57), I, 58–67. A concise account of the origins of Royal Dutch is given in Williamson *et al.*, I, 668–73, and Beaton, pp. 20–34.

[27]Gerretson, I, 67–127.

[28]Ibid., I, 127–62. [29]Ibid., II, 174–82.

and the process by which it came about clearly illustrates the major reason for many of the organization mergers in the early petroleum industry.

By the turn of the century a belief in the desirability of full vertical integration had become deeply entrenched in the petroleum industry, although most moves in that direction were perceived as solutions to specific, immediate problems.[30] Vertical integration meant bringing under a single management control of all phases of a petroleum business—production, refining, transportation, and distribution. Numerous advantages accrued, but the most obvious was the ability to ensure optimum utilization of expensive facilities by maintaining a constant flow through the entire system.[31] Sudden idling of a refinery when a non-owned distributor switched its source of supply, or inability to meet a distribution contract when a non-owned refinery decided to sell to a competitor could play havoc with profit margins. Rockefeller had clearly demonstrated the advantages when he built the original Standard Oil Trust by integrating forward from refining into transport and distribution and then backward into production. Competition frequently found itself forced to adopt the same tactic to survive, and this process was precisely what occurred in the case of Royal Dutch and Shell.

When Libby of Standard became convinced that petroleum from the Indies would soon become a major factor in the Asian trade, he made two abortive attempts to purchase Royal

[30]McLean and Haigh, p. 664.

[31]Ibid., pp. 663–77. A number of economic, technological, and managerial factors combined to favor vertical integration in the petroleum industry, but McLean and Haigh's inference that it was probably the optimum form of organization has been questioned though not disproven. A summary of some of the issues under debate is given in Harold F. Williamson and Ralph L. Andreano, "Integration and Competition in the Oil Industry." *Journal of Political Economy*, LXIX (August 1961), pp. 381–85. The economics of integration are extensively analyzed in Melvin G. deChazeau and Alfred E. Kahn, *Integration and Competition in the Petroleum Industry* (New Haven: Yale University Press, 1959).

Dutch—in 1895 and again in 1897.[32] Standard then tried in 1898 to acquire control of the Moreara Enim Petroleum Company, which had concessions in southern Sumatra. This time Standard was blocked by the Dutch government. Moreara indicated it would consider an agreement, but the Dutch Minister of Colonies hinted that if it did, the Sumatra concession might not be renewed, and that effectively ended negotiations.[33] When it became clear that penetration into the Indies had been effectively blocked, Standard turned to further intensification of its marketing campaign, including price wars in areas where Royal Dutch was attempting to gain a foothold.[34]

Shell's source of supply was threatened by rumors of Russian nationalization of its petroleum industry in the late 1890s, and Samuel had obtained a concession in Borneo, contracted for Moreara Enim's output, and established initial contacts in the new Texas oil fields, but dependable supply continued to be a concern.[35] Royal Dutch had moved into distribution throughout Asia but did not have a sufficiently broad base to withstand Standard's selective price cutting. As Deterding put it later, "to survive these price-cutting onslaughts we just had to hit back. . . [but] . . . if we hit back alone we would be knocked clean out of the fighting ring."[36] Standard thus helped

[32]Hidy and Hidy, pp. 263–65; Gerretson, I, 282–86. A desire on the part of Royal Dutch to retain its independence plus dissatisfaction over the financial terms offered by Standard appear to have been the primary reasons for both rejections.

[33]Peter M. Reed, "Standard Oil in Indonesia, 1898–1928," *Business History Review*, XXXII (Autumn 1958), pp. 312–13; Hidy and Hidy, pp. 265–66; Gerretson, II, 57–76. Reed infers that preliminary discussion of the Dutch mining law of May 23, 1899, permitting only Dutch citizens on the boards of companies operating in Indies oil fields directly influenced the decision of Moreara Enim to terminate negotiations. This appears doubtful since negotiations had collapsed a year earlier. Subsequent passage of the law does confirm, however, that the tide of nationalism was running high in the Netherlands.

[34]Hidy and Hidy, p. 267; Beaton, p. 46.

[35]Ibid., pp. 42–48.

[36]Sir Henri Deterding (as told to Stanley Naylor), *An International Oil Man* (London: Harper, 1934), p. 63.

create its own competition in the Asian oil business, and as it turned out, the world. Deterding first convinced Samuel and the Rothschilds to participate in a jointly owned marketing organization for the Orient, the Asiatic Petroleum Company, which began operations in 1903. "The companies forming Asiatic had, in effect, agreed to stop fighting each other and concentrate on Standard."[37] This worked so well that the unique Royal Dutch-Shell alliance was created in 1907. Royal Dutch and Shell were converted into holding companies, sharing ownership or a sixty-percent/forty-percent basis in three operating subsidiaries: Asiatic for marketing, N.V. De Bataafsche Petroleum Maatschappij (BPM) for production and refining, and the Anglo-Saxon Petroleum Company, Ltd., primarily for transportation.[38] When BPM acquired control of Dordtsche, the last significant independent producer in the Indies, in 1912,[39] Standard was faced with competition from a far-flung, fully-integrated organization in virtually complete control of Asia's most promising source of petroleum. It became exceedingly clear that if Standard were to remain in the Asian oil trade, it would have to acquire a source of supply in that region.

WHILE competition with Royal Dutch-Shell for the Asian kerosene market provided the original spark for Standard's interest in the Indies, two other factors added fuel to the fire. First, division of properties under court-ordered dissolution of the original Standard Oil combine in 1911 left the Standard Oil Company (New Jersey) with extensive refining and marketing facilities, but inadequate production sources for fully integrated operations. The company mounted a worldwide search for new sources to rectify this, and the Indies represented one

[37]Beaton, p. 49.

[38]Ibid., pp. 52–55. The Rothschilds retained an interest in Asiatic until the 1930s, but did not join in the rest of the Royal Dutch-Shell alliance.

[39]John S. Furnivall, *Netherlands India: A Study of Plural Economy* (Cambridge: Cambridge University Press, 1939), p. 328; Reed, p. 314.

25

of the most promising areas. Second, increasing military and civilian demand for petroleum products and concern over the adequacy of known petroleum reserves combined to cause the American government to provide maximum support for *any* American company searching for oil overseas in the early 1920s. As a result Jersey received ample diplomatic support in breaking down Dutch opposition, and became the second largest producer in the Indies.

The 1911 court dissolution of Standard Oil came as a blow to its executives, but in retrospect the decision should have surprised no one. Public, legislative, and judicial indignation had been mounting for years over stories of predatory price cutting, rebates, and industrial espionage, but in the final analysis it was Standard's sheer size that probably caused the most resentment.[40] Competition had begun to increase after 1900, but Standard still controlled roughly two-thirds of the American petroleum industry at the time the Supreme Court ordered it to divest itself of control over thirty-three subsidiaries.[41] The original Standard Oil Trust of 1882 had been converted into a loose holding company structure in 1892 with Standard Oil (New Jersey) at the center, and direct legal control of forty-one affiliated companies had been consolidated in the Jersey corporation in 1899.[42] It was this latter transaction that the Supreme Court declared illegal in 1911, and dissolution meant that Jersey (which remained the largest survivor) had to divest itself of a long list of subsidiaries—including Standard of New York, the Vacuum Oil Company, and Standard of California.[43] The recently established Asian distribution network went with Standard Oil of New York, which recombined with the Vacuum Oil Company in 1931 to form Socony-Vacuum (later renamed Mobil). California Standard later became a rival of Jersey in the Orient when in 1936 it linked with The Texas

[40]Hidy and Hidy, pp. 671–718.

[41]Williamson *et al.*, II, 4–14; Gibb and Knowlton, pp. 3–10.

[42]Hidy and Hidy, pp. 219–32; 306–13.

[43]Hidy and Hidy, pp. 697–98, 709–14; Gibb and Knowlton, pp. 3–10.

Company to form Caltex for overseas marketing operations. Of more immediate interest was the position Jersey itself occupied after the 1911 dissolution decree. The parent corporation had never been a producing company, and immediately after severance from most of its subsidiaries, Jersey could directly command a domestic production of only 7,500 barrels/day to meet a refinery through-put of 96,000 barrels/day.[44] By 1918 Jersey still produced only sixteen percent of the crude it processed, and this constituted one of the major problems faced by Walter C. Teagle when he assumed the presidency in 1917.[45]

By coincidence Teagle's drive to rebuild a fully integrated company coincided with an upsurge in American interest in foreign sources of petroleum. In 1904 a United States Navy Fuel Oil Board had recommended conversion from coal to oil, and within ten years the American fleet was well on its way to full conversion. By 1919 the Navy annually consumed almost 6 million barrels of fuel oil, a fraction of total American fuel oil consumption but highly significant from a strategic viewpoint.[46] A simultaneous surge in production of internal combustion engines for automobiles and trucks produced a rise in American gasoline consumption from approximately 6 million barrels in 1899 to over 87 million barrels in 1919.[47] The experience of mobilizing to meet American and Allied requirements in World War I appears to have had a profound psychological effect on oil experts both in and outside of government,

[44]Gibb and Knowlton, p. 44. These figures are for 1912 and represent the production of Standard Oil of Louisiana, Carter Oil, and a few other subsidiaries retained after the dissolution. The balance of Jersey's crude requirements were purchased, chiefly from former affiliates.

[45]Ibid., pp. 105–109, 278.

[46]Williamson et al., II, 181–84. The British Navy began a similar process of conversion to fuel oil on its smaller vessels about 1902 and became fully committed to oil in 1913. On American naval interest, see Nash, pp. 9–10, 16–20, 44–45, and John A. DeNovo, "Petroleum and the United States Navy Before World War I," Mississippi Valley Historical Review, XLI (March 1955), 641–56.

[47]Williamson et al., II, 184–95.

since they emerged with an almost universal concern over the adequacy of American resources. To this was added a series of highly pessimistic forecasts by the United States Geological Survey, especially a 1919 prediction that American petroleum reserves would be exhausted within ten years. All this led to mounting demands for a worldwide search for other sources of supply, and by mid-1919 the Department of State began instructing its officers overseas to give full support to this drive.[48]

While all three factors had worldwide significance, Jersey's desire to compete with Royal Dutch-Shell in its own backyard, Teagle's interest in rebuilding a fully integrated company, and the American "oil scare" after World War I converged with special force on the Netherlands East Indies. Some groundwork had already been laid. Dutch mining law permitted concessions to be granted only to Dutch citizens or corporations, so after failing to buy its way into the Indies by purchase of Royal Dutch, Moreara Enim, or Dordtsche, Jersey arranged for The American Petroleum Company, a marketing subsidiary in Holland that remained part of Jersey after the 1911 dissolution, to establish a new corporation under Dutch law to prospect in the Indies.[49] Thus was born in 1912 the Nederlandsche Koloniale Petroleum Maatschappij (or NKPM), originally a subsidiary of American Petroleum, and later the

[48]Nash, pp. 23–48; John A. Denovo, "The Movement for an Aggressive American Oil Policy Abroad, 1918–1920," *American Historical Review*, LXI (July 1956), 854–76. This "oil panic" lasted until about 1924, when new discoveries in Oklahoma, Texas, and California not only removed the threat of exhaustion but actually produced a surplus. Diplomatic support for American oil companies did not subside as quickly, however, and traces were still apparent years later; see testimony given by State Department Petroleum Advisor Charles Raynor before a Senate Committee in 1945; U.S. Congress, Senate, Special Committee Investigating Petroleum Resources, *American Petroleum Interests in Foreign Countries, Hearings*, 79th Congress, 1st Session (Part 3 of 6 parts; Washington, D.C.: Government Printing Office, 1964), hereafter cited as *American Petroleum Interests in Foreign Countries*, pp. 1–26, 297–324.

[49]Hidy and Hidy, p. 502; Reed, p. 315; Gould, p. 53.

producing component of the Standard-Vacuum Oil Company. NKPM began by purchasing a few unpromising concessions from other companies and attempting to obtain new permits from the Dutch government. This latter course was blocked in 1913 when the governor general of the Indies suspended the mining law and refused to grant new permits to anyone for an indefinite period.[50] Despite this discouraging prospect, NKPM began test drilling in 1912, struck oil in central Java in 1914, and tapped oil in what was later to become the Talang Akar field in southern Sumatra in 1916.[51]

None of the first wells proved commercially significant, and Jersey pressed the Dutch government for permission to explore the potentially rich Djambi fields in central Sumatra, enlisting State Department support in the process. The American Minister to The Hague, William Phillips, had repeatedly urged the Dutch Foreign Office to permit American participation in the Djambi fields, but the issue became entangled in a web of negotiations involving the Dutch Minister of Colonies, Royal Dutch-Shell, Sinclair Oil, The Texas Company, and Japanese interests, as well as Jersey. When the web was unraveled in 1921, the Djambi concession had gone to Royal Dutch-Shell under an arrangement whereby profits would be shared equally with the Dutch government.[52] Jersey's persistence wavered, but in 1922 deep drilling at Talang Akar finally tapped a commercially significant reservoir, and Teagle stoutly refused to consider suggestions from within the company that

[50]Gibb and Knowlton, p. 391; Reed, p. 315; Gould, p. 53. Suspension of the mining law appears to have been caused by inability to resolve a host of conflicting pressures—a distaste for foreigners, socialist opposition to private ownership of any kind, governmental interest in acquiring a share of the profits, Deterding's opposition to government participation, and a new interest in somehow ensuring a fuel supply for the Dutch Navy (Gerretson, II, 307–16, and IV, 92–98). A new mining law was finally passed in 1918.

[51]Gould, pp. 54–55.

[52]Reed, pp. 318–28; Gould, pp. 55–59; Gibb and Knowlton, p. 392. The issue was resolved through a law passed in the Netherlands Estates General in April 1921, with the support of the Minister of Colonies.

Jersey sell out to Socony or the Japanese.[53] By the end of 1926 twenty-four producing wells had been brought in at Talang Akar, a six-inch pipeline had been laid through eighty miles of jungle, and a 3,500 barrel/day refinery had been built on landfill and pilings in the swampy land at Soengi Gerong, ten miles south of Palembang in southern Sumatra.[54] Jersey had a firm toehold, but its concessions in the Indies still amounted to only sixty-eight thousand acres compared with Deterding's control over five million acres.[55]

Parallel with this development, pressure had been mounting in the United States to increase governmental support for Jersey. Largely under the influence of the post-World War I oil scare, Congress included in its 1920 Mineral Leasing Act a provision that no leases on public lands be permitted for foreign corporations whose parent countries did not grant similar privileges to American interests. Partly as a result of the Djambi affair, Secretary of the Interior Albert Fall in 1922 designated the Netherlands as a "non-reciprocating country," an action with

[53]Ibid., p. 393. Legend has it that a cable from Djakarta ordering the drilling crew to abandon an apparently dry hole was delayed by the Christmas holiday, and the crew struck oil on December 26, 1921. The story is plausible, but could be traced to only one undocumented source, the National Planning Association's *Stanvac in Indonesia: Sixth Case Study in an NPA Series on United States Business Performance Abroad* (Washington, D.C.: National Planning Association, 1957), hereafter cited as *Stanvac in Indonesia*, p. 22. Jersey's historians give the date of the Talang Akar breakthrough as August 1922 (Gibb and Knowlton, p. 363). In 1940 Stanvac Board Chairman George Walden cited Talang Akar No. 6, "completed in December 1922" as "the first producing well in what proved to be the main producing sand" in that field (letter, G. S. Walden to W. C. Teagle, April 4, 1940, files of the Exxon Corporation in New York, hereafter cited as "Exxon files"); this is one of a few documents relating to Stanvac operations located; it was made available to the author through the courtesy of Exxon's New York office.

[54]Gibb and Knowlton, 393; Gould, pp. 61–63; *Stanvac in Indonesia*, p. 23; *Standard-Vacuum Petroleum Maatschappij: Forty Years of Progress, 1912–1952* (Djakarta, Indonesia: Standard-Vacuum Petroleum Maatschappij, n.d.), pp. 6, 26, copy in the library of the Exxon Corporation in New York, hereafter cited as "Exxon library."

[55]Gibb and Knowlton, p. 393.

more psychological than practical effect since it only denied Shell access to public lands in the United States.[56] The label of "non-reciprocity" was used quite effectively, however, by the new American Minister to the Netherlands, Richard Tobin, to keep sustained pressure on the Dutch government to grant more extensive concessions to Jersey in the Indies. In 1923 Jersey began negotiations for an extension of its Talang Akar concession under a so-called "5A contract" granting the Dutch government twenty percent of the profits, but this required approval of the Estates General, and the issue remained unresolved for five years. After sustained diplomatic exchanges, the Estates General finally approved the transaction on July 17, 1928, Jersey acquired a concession for 625,692 additional acres of promising land, and the American government removed the "non-reciprocity" label from the Netherlands.[57] Years later a United States Senate Committee concluded that ". . . it is very doubtful that Americans could have entered. . . [the Netherlands Indies] . . . on any appreciable scale . . ." without governmental support.[58] Not only had an American company become solidly entrenched along with Royal Dutch-Shell in the Indies, but a precedent of cooperation on East

[56]Williamson *et al.*, II, 518; Gould, pp. 55–56, 60; Reed, p. 328. In 1923 Fall also denied a Shell subsidiary, Roxana, permission to lease certain Indian lands, but this decision was promptly recinded by Fall's successor. Shell's historian suggests that Fall's action against Shell may have been an attempt to draw attention away from the developing Teapot Dome scandal which shortly thereafter forced his resignation (Beaton, pp. 230–34).

[57]Reed, pp. 329–36; Gould, pp. 60–62; Gibb and Knowlton, pp. 393–94.

[58]*American Petroleum Interests in Foreign Countries,* p. 323. For a contrary view see Nash, pp. 60–62; Nash did not consider the 1928 concession of great importance and therefore concluded that American diplomatic efforts had been inept and ineffective. Nash appears to have overlooked the fact that as a result of the change in Dutch concession policy NKPM was able to increase its Indies production from approximately one million barrels in 1927 to over sixteen million barrels in 1937, and Stanvac was able to meet half its total marketing requirements from production in the Indies by the late 1930s; see *SVPM: Forty Years of Progress,* pp. 22–23, Exxon library; data in Appendix B; and Larson, Knowlton, and Popple, p. 147.

Asian matters had been firmly established between Jersey and the American State Department.

BY THE end of the decade, a new climate had begun to permeate the international oil business. Fear of shortage had passed, and by 1928 new discoveries began to create a surplus large enough to depress world prices.[59] A disagreement between Standard of New York and Royal Dutch-Shell over how to counter aggressive Russian tactics sparked a price war in India which spread to other continents and convinced many oil men that hard-won distribution systems could be lost overnight if such competitive practices continued unabated.[60] And finally there was the need to settle a long-standing quarrel over how the promising new Middle Eastern fields were to be developed. The time was ripe for compromise. Largely at the prompting of Deterding and Teagle a number of international firms reached agreements dividing up Middle Eastern oil and most of the world market outside the United States in order to reduce the risk of "boom or bust" production and costly price wars. A spirit of cooperation dominated the major firms well into the 1930s, and this was the world into which the Standard-Vacuum Oil Company was born. Stanvac, which proved to be an efficient combination of Socony-Vacuum's Asian distribution system with Jersey's new production and refining operation in the Indies, was a fully integrated and relatively independent joint subsidiary. By 1933 fifty years of alternating competition and compromise had spawned an organization that represented the single largest American direct investment in East Asia and the link pin in an informal network controlling the lion's share of the Asian oil trade. At birth Stanvac inherited a close working relationship with Royal Dutch-Shell and a precedent of cooperation with the American State Department.

The incredibly complex story of Middle Eastern oil is well

[59]Williamson *et al.*, II, 524–25, 532; Gibb and Knowlton, pp. 304–305; Larson, Knowlton, and Popple, pp. 305–306.

[60]Williamson *et al.*, II, 527–30; Larson, Knowlton, and Popple, pp. 305–306.

beyond the scope of this chapter, but a few salient points are worth noting.[61] British and German interests had begun investigating concessions in geologically promising areas prior to World War I, and the governments of Britain and France had attempted to divide access to the region between themselves through the ill-fated San Remo Oil Agreement of 1920. This was at the height of the postwar oil scare, and the American government registered forceful protests over the exclusion of American interests—including representations made at the 1923 Lausanne Conference by the chief American delegate, Joseph C. Grew. Resolution of the issue dragged on until July 31, 1928, when all parties finally agreed to a plan for joint ownership of the Turkish Petroleum Company and signed the famous "Red Line Agreement."[62] Its name derived from a red line drawn on a French map outlining the old Ottoman Empire,

[61]A clear and well-researched account of this episode is given in Gibb and Knowlton, pp. 278–316. A slightly more critical version is contained in U.S. Federal Trade Commission, *The International Petroleum Cartel*, FTC Staff Report Submitted to the Subcommittee on Monopoly of the Select Committee on Small Business, U.S. Senate; 82nd Cong., 2nd Sess. (Washington, D.C.: Government Printing Office, 1952), hereafter cited as *"International Petroleum Cartel,"* pp. 47–67.

[62]The complexity of the negotiations may be deduced from the final distribution of shares in the Turkish Petroleum Company in 1928, as extracted from Gibb and Knowlton, pp. 283, 306, and *International Petroleum Cartel*, pp. 49, 65:

23¾ % – D'Arcy Exploration Co., Ltd. (Subsidiary of the Anglo-Persian Oil Company, in which the British government held a 56% interest.)

23¾ % – Anglo-Saxon Petroleum Co., Ltd. (Subsidiary of Royal Dutch-Shell, held 60% by Royal Dutch and 40% by Shell Transport and Trading.)

23¾ % – Compagnie Française des Petroles (Organized in 1920 by the French government to take over the former holdings of the Deutsche Bank in the Turkish Petroleum Company.)

23¾ % – Near East Development Corporation (Organized in 1928 for participation in TPC by five American companies, with shares distributed as follows:

25%   – Standard Oil Company [New Jersey]

and the agreement essentially bound all participants to acquire crude only through TPC within the area that later became Turkey, Syria, Lebanon, Israel, Jordan, Saudi Arabia, and Iraq.[63] This understanding has been called both "one of the outstanding instances of international sharing and cooperation" and "an outstanding example of a restrictive combination for the control of a large portion of the world's oil supply."[64] Regardless of one's viewpoint, it clearly represented a milestone in the development of cooperative relationships between Jersey and Royal Dutch-Shell.

Six weeks later Teagle, Deterding, and Sir John Cadman of Anglo-Persian worked out a second pact, known as the "Achnacarry Agreement" for the Scottish castle where discussions took place. Marketing agreements for specific areas were by no means new to the international petroleum industry, but they generally had been narrow in scope and short lived. New York Standard had set an early precedent by agreement on prices and percentages of the China market with Shell in 1908, but this lasted only two years. A resurgence of intense competition between these two firms in the Indian market in 1927 sparked an international price war which set the stage for Achnacarry.[65] The experience of the price war coupled with

---

25% – Standard Oil Company of New York

16⅔% – Gulf Oil Corporation

16⅔% – The Atlantic Refining Company

16⅔% – Pan American Petroleum and Transport Company.)

5% Participation and Investments Company (Wholly owned by C. S. Gulbenkian, "the Tallyrand of Middle Eastern Oil politics.")

[63]Gibb and Knowlton, pp. 306–308; *International Petroleum Cartel,* pp. 60–67. Kuwait and Iran were not included; TPC changed its name to Iraq Petroleum Company in 1929.

[64]Ibid., p. 67, quoting studies by C. S. Morgan in the first instance and R. F. Mikesell in the second.

[65]Hidy and Hidy, pp. 549, 553; Gibb and Knowlton, pp. 79, 201; Larson, Knowlton, and Popple, pp. 305–306; Williamson *et al.,* II, 258-59, 528–30; *International Petroleum Cartel,* pp. 197–98.

mounting concern for overproduction prompted Teagle, Deterding, and Cadman in September 1928 to attempt an agreement allocating the world market outside the United States on the basis of proportionate shares as they existed in that year.[66] The document was never formally signed, but it provided the basis for a long series of attempts at regional agreements during the early 1930s. One specific byproduct of this movement was a new agreement in 1930 between Royal Dutch-Shell and Standard of New York on pricing in the Orient.[67] Thus by the early 1930s a strong precedent of cooperation with Royal Dutch-Shell had been established by both Jersey and Standard of New York (Socony-Vacuum after 1931).

The final link in what proved to be a highly effective network was forged in 1933 with formation of the Standard-Vacuum Oil Company itself. By this time overproduction and economic depression had created a general trend toward cooperation to minimize inefficiency and wasteful competition,[68] but the initial impulse for Stanvac came from an altogether different source.[69] As already noted, NKPM's 1928 concession in the Indies allocated twenty percent of the net profit to the Dutch government, and discontent developed over interpretation of

[66]Williamson *et al.*, II, 530; Larson, Knowlton, and Popple, pp. 308–309; *International Petroleum Cartel*, pp. 199–210. The "agreement" specifically excluded the United States market presumably to avoid conflict with American antitrust legislation, and was considered to be completely legal at the time. Combinations of American firms in foreign trade were permitted under the Webb-Pomerene Act of 1918, provided they did not also work to produce a restraint of trade within the United States. *It was the interpretation of this clause that subsequently caused legal difficulties.* Although Standard of New York had helped start the price war, it did not participate in the Achnacarry Agreement.

[67]Williamson *et al.*, II, 530–32; Larson, Knowlton, and Popple, pp. 309–13; *International Petroleum Cartel*, pp. 211–74.

[68]Williamson *et al.*, II, 540–50; Nash, pp. 98–126; Larson, Knowlton, and Popple, pp. 303–14.

[69]This account of the business reasons for the formation of Stanvac is based on Larson, Knowlton, and Popple, pp. 147, 315–18; and Williamson *et al.*, II, 529.

this provision. In 1927 Jersey had contracted with Standard of New York to market NKPM's entire output, a logical arrangement since Jersey had no Asian distribution system, but one which deprived NKPM (and the Dutch) of additional profits it would have earned had NKPM handled its own marketing. This scheme sorely displeased the Dutch, who began to hint that additional concessions might not be forthcoming unless some more profitable arrangement could be worked out. Jersey was thus pressured into seeking another solution to its marketing problem.

Developing its own Oriental distribution system would have been a costly proposition, so the company broached the idea of an Asian merger to Socony-Vacuum, which was initially cool. The possibility of a merger with Royal Dutch-Shell was considered, but Jersey then decided to attempt to purchase from Atlantic Refining and Union Oil their joint marketing subsidiary in Australia and New Zealand. At this point Socony had a change of heart, apparently caused by its interest in finding a more dependable source of supply than the Russian crude it had come to rely upon for its Asian market. After some hard negotiating Socony agreed with Jersey to form a new joint subsidiary—the Standard-Vacuum Oil Company, organized September 7, 1933, as a Delaware corporation.[70] Each parent retained a fifty-percent interest, and Socony-Vacuum's entire distribution network south and east of Suez and Jersey's entire operation in the Indies were transferred to the new corporation. Half of Stanvac's officers and directors came from Jersey and half from Socony-Vacuum,[71] and they were given a relatively free hand in running what was now essentially a self-contained operation.

Much of Stanvac's initial success stemmed from the extra-

[70]*Moody's Industrial Manual* (New York: Moody's Investment Service, 1939), p. 3088.

[71]List of Stanvac directors and officers provided to the author by Charles E. Springhorn of the Exxon Corporation by letter of April 25, 1969.

ordinary compatibility and enthusiasm of its key personnel.[72] The final organizational details had been hammered out by Edward J. Sadler of Jersey and William B. ("Billy") Walker of Socony. As Jersey's top production man for over a decade, Sadler knew his side of the business thoroughly, and Walker had equal knowledge of Asian marketing from years of ranging up and down the China coast. The two men worked well together, and they developed a corporate structure that lasted almost thirty years. Walker himself became Stanvac's first president, but on his sudden death in 1934 a new management arrangement was inaugurated. George S. Walden, head of Jersey's NKPM operation since the late 1920s, became board chairman and chief executive officer, and the presidency went to Philo W. Parker, a veteran of twenty years in China marketing with Socony. Possessed of a shrewd business sense and exceptionally easy to talk with, Walden had originally joined Standard in 1908 as an office boy. After a two-year tour in France in World War I, he entered the production side of the business and worked his way up from roustabout to corporate executive with assignments in West Virginia, Oklahoma, Argentina, the Indies, and New York, developing a personal friendship with Teagle and Sadler in the process. Tough minded and tenacious, Walden had begun Jersey's original construction at Palembang and then spent four years commuting between Batavia, The Hague, and New York to solidify organization of the Dutch-chartered NKPM. While carefully preserving his authority as chief executive officer, Walden permitted his teammate, Philo Parker, to take the initiative in matters affecting marketing outlets in China and Japan, where

---

[72]This account of Stanvac's original management team is based on Larson, Knowlton, and Popple, p. 339; brief biographical sketches in *The Lamp*, February 1934, pp. 20–27, Exxon library; an interview with Lloyd W. Elliott, managing director of NKPM in the late 1930s, in New York on August 25, 1970; and an interview with Edward F. Johnson, one of Stanvac's original vice presidents, in New York on August 11, 1971.

Parker was far more experienced. Well educated, reserved, and exceptionally diplomatic, Parker complemented Walden perfectly. He had joined the old Standard Oil of New York organization in 1912 for the purpose of entering foreign sales, and an initial training assignment in Hong Kong had led to twenty years in China before returning to New York in 1932. In the process he had become thoroughly familiar with the intricacies of Asian business and politics, and had developed considerable patience and a keen sense of timing. Together with a seasoned team of oil men from both parent companies, these two men headed a far-flung, aggressive operation which was about to feel the pressure of Japanese expansion.

THE Asian oil trade had come a long way since the days of wooden packing cases and kerosene lamps. Standard had begun with sales through commission merchants, but converted to establishing its own distribution system in response to competition from Russian kerosene in the 1890s. Pressure from Standard had in turn helped create the vast Royal Dutch-Shell group with a strong base in Indies petroleum. Reaction to this development, coupled with a desire to rebuild an integrated company after the 1911 dissolution, and a general American concern over supply after World War I had combined by 1928 to prompt Jersey to acquire a significant share in Indies production, and to establish precedents for collaboration with the Department of State in the process. The same year saw a turn toward cooperation between Jersey, Standard of New York, and Royal Dutch-Shell, symbolized by the Middle Eastern Red Line Agreement and the informal Achnacarry pact. A further linkage came in 1933 when Jersey and Socony-Vacuum merged their Oriental interests to form the Standard-Vacuum Oil Company. An embryonic network had thus emerged which events of the 1930s were to forge into a tightly knit team.

38

THE PRINCIPLE OF THE OPEN DOOR

DURING THE 1930s, problems in China and the specter of Japanese expansion accelerated trends toward closer cooperation between Stanvac, Shell, and the Anglo-American diplomatic corps, especially in cases that involved defense of traditional Open Door treaty rights. From a business point of view, the problems faced by Walden and Parker were straightforward, but the entanglement of commercial issues with treaty rights created a complex situation, and careful analysis of the period sheds considerable light on how different but intersecting institutional interests could prompt businessmen and diplomats to develop an informal but smooth-working defensive team.

In Canton, along with Asiatic Petroleum and The Texas Company, the new corporation faced a maze of discriminatory local regulations complicating the sale of kerosene and requiring a solidly united front and a full year of negotiations to overcome. In Manchuria, the Japanese-controlled Manchukuo government in 1934 formed a petroleum monopoly which, despite strong official objections from Britain and the United States, finally forced Stanvac, Asiatic, and The Texas Company to completely withdraw from that area. Ultimately of more significance, a 1934 Petroleum Industry Law in Japan established extensive control over the industry and required all distributors of petroleum products to build up and maintain a six-month stockpile at their own expense. Walden and Parker objected vigorously to this provision and, along with Rising Sun Petroleum (Asiatic's subsidiary in Japan), mounted a prolonged, intricate, and ultimately successful campaign to negotiate their way out of compliance. All three episodes prompted diplomatic intervention, and State Department and

Foreign Office records reveal the evolution of a highly effective working relationship as a result of this experience.

The business reasons for opposing such moves were clear, but American diplomats were plagued with two almost impossible dilemmas, both deeply imbedded in a summary of policy written in December 1933 by Nelson T. Johnson, the genial and articulate American Minister to China. Recording a conversation with the Assistant Chief of the State Department's Far Eastern Division, then visiting in Nanking, Johnson stated that:

> The United States is interested merely in seeing that American citizens get a fair share of China's market for their products. To obtain this share we are not interested either in taking Chinese territory or in extending a protectorate over any part of China. On the contrary we are interested in seeing China grow strong and able to defend herself within her boundaries, believing that a strong China able to fend for itself is not only the best guarantee to Americans for a fair share of any demand that might arise in China for foreign products, but also the best guarantee of peace in the Far East. It is my belief that it is our desire to enjoy the friendliest relations, not only with the Chinese, but with the Japanese. But I feel that in order that we may enjoy these friendly relations it is necessary that we give ample evidence that we are prepared to defend our rights, wherever those rights may exist, against attack or encroachment from any source.[1]

Carefully read, this version of America's traditional Open Door policy left something to be desired in intellectual rigor, but it did contain a concise summary of the contradictory impulses deeply ingrained in the American diplomatic establishment of

[1]Memorandum of conversation with Maxwell M. Hamilton, in Nanking, December 10, 1933; Nelson T. Johnson MSS, Volume 36, pp. 300–301, Library of Congress, Washington, D.C., hereafter cited as "Johnson MSS."

1933.[2] Much internal agonizing in the 1930s centered on attempts to resolve the twin dilemmas of how to maintain friendship with both Japan *and* China and how to "defend our rights" in an area not perceived as sufficiently vital to the national interest to warrant permanent commitment of major military force. The task became almost impossible when American diplomats were also confronted with the ambiguities of Asian politics which defied reduction into neat Western formulae of right and wrong. More often than not, American diplomats resolved the problem by reverting to legally correct positions and couching them in language carefully designed to antagonize no one. Inevitably, this procedure drew Americans

---

[2]A full account of the Open Door policy would be almost synonymous with the complete story of Sino-American relations in the first half of the twentieth century. Among the many works touching on Secretary of State John Hay's original Open Door notes of 1899 and 1900, four are especially useful: A. Whitney Griswold, *The Far Eastern Policy of the United States* (New York: Harcourt, Brace & World, 1938); Paul A. Varg, *Open Door Diplomat: The Life of W. W. Rockhill* (Urbana: University of Illinois Press, 1952); Charles S. Campbell, Jr., *Special Business Interests and the Open Door Policy* (New Haven: Yale University Press, 1951); and Marilyn B. Young, *The Rhetoric of Empire: American China Policy, 1895–1901* (Cambridge: Harvard University Press, 1968). For a sound recent account of the Washington Conference, where the Open Door policy was written into the Nine-Power Treaty of 1922, see Thomas H. Buckley, *The United States and the Washington Conference, 1921–1922* (Knoxville: University of Tennessee Press, 1970). By far the most useful study of American Asian policy in the mid-1930s is Dorothy Borg's *The United States and the Far Eastern Crisis of 1933–1938* which concludes that despite the tradition of the Open Door, American diplomats studiously tried to avoid antagonizing Japan. For two contary views, at opposite ends of the ideological spectrum, see George F. Kennan, *American Diplomacy, 1900–1950* (Chicago: University of Chicago Press, 1951) arguing that American policy became excessively legalistic and moralistic; and William A. Williams, *The Tragedy of American Diplomacy* (Rev. ed.; New York: Dell, 1962) suggesting that the driving force stemmed from a sophisticated form of economic imperialism.

into increasing diplomatic opposition to Japan, since that was the country expanding its sphere of influence in technical violation of several treaties dear to the United States, but the ultimate consequences of such a trend were not fully discerned in 1933. From the point of view of Walden and Parker, this posture meant that Stanvac could count on diplomatic support only when business and legal interests coincided, but it was Japanese infringement on precisely these two domains that produced the close teamwork between British and American businessmen and diplomats in the early 1930s.

ALTHOUGH Walden and Parker contributed heavily to the maturing of this pattern of cooperation, neither initiated it; it was already in operation at lower echelons of the organization even before the consolidation of Socony and Jersey interests in Asia. A troublesome "kerosene war" was in progress in Canton at the time Stanvac was created, and careful analysis of the details of that episode reveals a pattern that was to become amplified many times over as the decade progressed.

Before recounting the story, a brief note is in order on the political climate in Canton in 1933 and 1934. Casual observers of Chinese history sometimes fail to note that Chiang Kai-shek's Nationalist government never completely consolidated its authority over all of China in the 1930s. Campaigns against recalcitrant Communist enclaves made headlines but represented only one of many problems. Dissident generals plagued Chiang with repeated attempts to retain or establish local autonomy, and the Canton "Southwest Political Council" supported by Generals Ch'en Chi-t'ang and Li Tsung-jen was one such effort. Li had participated in the Kwangsi generals' revolt of 1929, and he combined with Ch'en to oppose Chiang again in 1936. The national government, however, managed to co-opt the forces of Li and Ch'en by a skillful blend of placating the leaders and encouraging defection in the ranks. During much of the intervening period, the Southwest Political Council maintained effective control in Kwangtung and Kwangsi provinces, nominally as a branch of the central government, but actually demonstrating considerable autonomy and insubordination.[3] It was this autonomy that made the kerosene problem so difficult to resolve through normal diplomatic channels.

The Canton kerosene episode actually began in the fall of 1932, before the formation of Stanvac, and the issue continued to plague the new company until a complicated settlement was arrived at in September 1934. The "war" started when a group of Chinese entrepreneurs, allegedly including General Ch'en, some members of the Southwest Political Council, and the Chairman of the Kwangtung Provincial Government, decided to take advantage of a differential in the import duties on refined and crude oil of approximately $3.00 per unit of two tins (ten U.S. gallons). In late 1932 Chinese merchants began to import a grade of oil that passed customs as crude, distilling this into kerosene by a simple and inexpensive process. Since this kerosene was competing with foreign kerosene entering under the higher rate of duty, the Chinese merchants pocketed a substantial profit. The central government at Nanking immediately perceived a loss of customs revenue and directed that the duty on crude be increased to equal that on refined oil, but the Kwangtung government in Canton refused to allow the Customs Service to collect the increased tax on the ground that it would damage a native industry, namely distilling. When it became clear that Nanking could not prevail, the war was on, and the three foreign oil companies operating in South China launched a counterattack of their own. Socony-Vacuum, Asiatic Petroleum, and The Texas Company commenced importing the lower grade of oil, established their own local distilling

[3]Immanuel C. Y. Hsü, *The Rise of Modern China* (New York: Oxford University Press, 1970), pp. 633–40, 649–61; and Ch'ien Tuansheng, *The Government and Politics of China* (Cambridge: Harvard University Press, 1950), pp. 101–106, 364. Li Tsung-jen went on to become Vice President and presided over the final collapse of the Nationalists on the mainland as Acting President when Chiang fled to Taiwan in 1949.

operations, and began to flood the market, selling at a loss to drive their competitors out of business.[4]

A number of Chinese distillers withdrew from the field, but the remainder were saved by a new set of regulations issued by the Cantonese authorities in mid-1933. On July 3, the local Commissioner of Customs informed Joseph W. Ballantine, the American Consul General at Canton, and his British counterpart that the Southwest Political Council had decreed that special permits would be required to import the kerosene distillate. Separate regulations specified that permits could be issued only to "registered" distillers, and yet another set of regulations permitted only "Chinese-controlled" firms to register with the Cantonese government. Ballantine and the British Consul General promptly lodged protests against this obvious attempt to legislate foreign oil companies out of business, invoking, with the concurrence of their respective legations in Peiping, Article III of the Sino-American Commercial Treaty of 1903. By the time Socony-Vacuum's Hong Kong Office cabled New York and P. J. Gallagher of Socony's export department went down to Washington on July 21 to enlist the support of the Department of State, Far Eastern Division Chief and veteran Asian specialist Stanley K. Hornbeck could report that the American government was already actively at work, and the Department shortly thereafter cabled Johnson to in-

[4]Memorandum on "The Foreign Oil Companies and the Oil Subsidy at Canton" prepared June 6, 1934, by Ronald A. Hall of the British Foreign Office Far Eastern Department, British Foreign Office Political Correspondence, Class F.O.371, 1934, Public Record Office, London, hereafter cited as "FO371 (year)", registry number F3753/75/10; letter, Joseph W. Ballantine (American Consul General at Canton) to Nelson T. Johnson, July 10, 1933, Record Group 59, General Records of the Department of State, National Archives, Washington, D.C., hereafter cited as "RG 59," file 693.116/73; and memorandum of conversation, Peck (Counselor of the American Legation in China) with T. V. Soong (Chinese Minister of Finance) in Nanking, October 3, 1933, U.S. Department of State, *Foreign Relations of the United States* (Washington, D.C.: Government Printing Office, 1852–), hereafter cited as *"FRUS(year),"* 1933, III, 586–87.

clude Article XV of the Sino-American Treaty of 1858 in future protests. Treaty violations had spurred the diplomatic establishment to action with minimal prodding from the commercial interests involved.[5]

Steady diplomatic pressure at Canton, at Nanking, and at the Chinese Legation in Washington finally produced on August 18 a "clarification" of the registration regulations from the Southwest Political Council, stating that the requirement for "Chinese control" only applied to Sino-foreign firms, and purely foreign companies could of course register by completing a simple form. When Socony-Vacuum, The Texas Company, and Asiatic Petroleum at last obtained copies of this form they discovered that completion would, in effect, nullify their extraterritorial rights. The British and American Consuls General in Canton promptly advised the companies not to submit, and another round of diplomatic discussion ensued.[6]

---

[5]Telegram, Johnson to SecState, July 12, 1933, *FRUS(1933)*, III, 569–70; telegram, Phillips (Acting Secretary of State) to Johnson, July 28, 1933, *FRUS(1933)*, III, 575; memorandum of June 6, 1934, by R. A. Hall, FO371(1934)F3753/75/10; memorandum of conversation, Hornbeck with Gallagher, July 21, 1933, RG 59, 693.116/70. For the Sino-American Treaty of 1903, see *FRUS(1903)*, p. 91; for the Sino-American Treaty of 1858, see Hunter Miller, ed., *Treaties and Other International Acts of the United States of America* (8 vols.; Washington, D.C.: Government Printing Office, 1931–1948), VII, 793. The articles cited guaranteed the right of American citizens to engage in trade in China.

[6]Memorandum of June 6, 1934, by R. A. Hall, FO371(1934)F3753/75/10; letter, Ballantine to Johnson, August 30, 1933, Johnson MSS, Vol. 17, "Correspondence. 1933"; telegram, Carr (Acting Secretary of State) to Johnson July 22, 1933, *FRUS(1933)*, III, 573; telegram, Johnson to SecState, July 24, 1933, *FRUS(1933)*, III, 573; telegram, Johnson to SecState, July 27, 1933, *FRUS(1933)*, III, 574; telegram, Johnson to SecState, August 9, 1933, *FRUS(1933)*, III, 576; note delivered to the Chinese Minister in Washington, August 14, 1933, *FRUS(1933)*, III, 577–78; telegram, Johnson to SecState, August 15, 1933, *FRUS(1933)*, III, 578; telegram, Johnson to SecState, August 21, 1933, *FRUS(1933)*, III, 579; and telegram, Johnson to SecState, August 31, 1933, *FRUS(1933)*, III, 580–81. Extraterritoriality involved the exemption of foreign nationals from Chinese civil and criminal

Asiatic toyed with the idea of settling by going into partnership with the Chinese interests, but this was opposed by the British Consul General, who admitted that he was "endeavoring to fight for the principle of treaty rights as much as anything else."[7] Principle prevailed, and this round was settled on November 11 with an exchange of notes permitting the companies to register after the American and British Consulates General had extracted a pledge from Cantonese authorities that they had "no idea of proclaiming or enforcing laws and regulations governing oil factories which are in contravention of treaties in force."[8] The companies could once again legally import kerosene distillate.

Treaty rights might be intact, but the problem was far from solved. An intricate and ingenious new obstacle arose, requiring considerable patience to clear away. As early as August 17 Gallagher had alerted Hornbeck to a rumor that Canton might impose a local kerosene "business tax" of $3.00 per unit of ten gallons, with a private return of this tax to the Chinese distillers.[9] Such a tax did materialize in December, technically applicable to all retailers (to avoid the charge of a multiple transit tax on imports, or "likin," which was prohibited by treaty), and payable one-third in cash and two-thirds in promissory notes (which were admittedly cancelled in the case of Chinese-distilled kerosene). This ingenious system, outside

---

law, providing instead for prosecution in their own consular courts. The system originated with the "unequal treaties" of 1842–60, and was not relinquished by the United States until 1943. Several sections of the forms in question implied submission to Chinese law, and this was the sticking point. For a concise summary of the overall question of extraterritoriality, see John K. Fairbank, *The United States and China* (3rd ed.; Cambridge: Harvard University Press, 1971), pp. 142–49, 178, 203, 217–18, 305.

[7]Telegram, Johnson to SecState, September 25, 1933, RG59, 693.116/94.

[8]Memorandum of June 6, 1934, by R. A. Hall, FO371(1934) F3753/75/10; telegram, Gauss (for the Minister in China) to SecState, November 8, 1933, *FRUS(1933)*, III, 592; telegram, Gauss to SecState, November 11, 1933, *FRUS(1933)*, III, 592–93.

[9]Letter, Gallagher to Hornbeck, August 17, 1933, RG 59, 693.116/76.

the technical reach of treaty rights, permitted the Chinese distillers a competitive advantage of $2.00 per unit, or approximately twenty-five percent, and circumvented the Customs Service, creating revenue for the Kwangtung government that normally would have gone to Nanking.[10] Socony-Vacuum, Asiatic, and The Texas Company began to search for a counter weapon.

Socony's Hong Kong office suggested direct retaliation against Cantonese exports to the United States, and the State Department asked Johnson to consider seriously the feasibility of such action. Ballantine, however, pointed to the danger of triggering a general Chinese boycott, and H. L. Schultz, Hong Kong manager for what by then was Stanvac, cooled to the idea. The Division of Far Eastern Affairs finally tabled the proposal in April of 1934, and it passed quietly from the scene.[11] Stanvac, Asiatic, and The Texas Company attempted several compromise settlements but fell back on their ultimate weapon of selling at a loss to force out the competition. By mid-1934 this

[10]Memorandum of June 6, 1934, by R. A. Hall, FO371(1934) F3753/75/10; letter, Gallagher to Hornbeck, January 30, 1934, RG 59, 693.116/123; telegram, Johnson to SecState, October 17, 1933, *FRUS(1933)*, III, 588–89; telegram, Johnson to SecState, January 3, 1934, *FRUS(1934)*, III, 593; telegram, Johnson to SecState, January 31, 1934, *FRUS(1934)*, III, 565. No price for kerosene in Canton in 1934 could be located, but the retail price in Shanghai in the summer of 1935 stood at $7.90 "Chinese currency" per unit of ten American gallons of first grade kerosene, and it has been assumed that the 1934 Canton price would not have been radically different (Economic and Trade Note, "Reduction in Kerosene Prices," submitted by A. Viola Smith [American Trade Commissioner in Shanghai], September 3, 1935, Record Group 151, Records of the Bureau of Foreign and Domestic Commerce, National Archives, Washington, D.C., hereafter cited as "RG 151", Box 1521, Folder: "Peiping").

[11]Letters, Gallagher to Hornbeck, October 26, 1933, RG 59, 693.116/115; telegram, Hull to Johnson, November 4, 1933, *FRUS(1933)*, III, 590; letter, Johnson to SecState, February 10, 1934, RG 59, 693.116/127; letter, Schultz to Johnson, February 1, 1934, RG 59, 693.116/127; Division of Far Eastern Affairs memorandum of April 4, 1934, RG 59, 693.116/127.

tactic had begun to work, or at least to hurt the Chinese distillers badly, and a private settlement appeared possible.[12]

Peace came on September 28, 1934, after prolonged negotiation with the Kwangtung Department of Finance along lines suggested by J. W. Platt, Canton manager for Asiatic Petroleum. In a formal document of agreement, Stanvac, Asiatic Petroleum, and The Texas Company agreed to pay the Kwangtung Finance Department $5,000,000 in Hong Kong currency as an advance against the Provincial Kerosene Business Tax, in return for which the Finance Department agreed to tax all importers equally at the rate of $3.00 per unit and terminate all refunds or subsidies to Chinese distillers. Future tax assessments were to be deducted from the advance at a carefully prescribed rate, and the companies privately expected to recover their investment in a minimum of fourteen months. In effect, the oil companies had used price cutting to force an agreement, and then purchased equality of treatment by payment of $5,000,000 to the Kwangtung government against a tax that probably should have gone to Nanking. The *pièce de resistance* came, however, when the Chinese distillers organized a demonstration to protest the end of their subsidy. The manager of a Chinese distillery that had been secretly purchased by Asiatic asked Platt whether he could participate, and Platt not only gave permission but provided the funds to hire thirty more demonstrators. The newspapers carried no comment on the $5,000,000, and a restrained protest against the government's acquiescence to foreign pressure satisfied the proprieties of Cantonese society.[13]

This episode has been recounted in some detail less for its

[12]Memorandum of June 6, 1934, by R. A. Hall, FO371(1934) F3753/75/10; telegram, Johnson to SecState, June 14, 1934, *FRUS (1934)*, III, 566.

[13]Letter, Herbert Phillips (British Consul General at Canton) to Sir Alexander Cadogan (British Minister in China), October 11, 1934, enclosing the full text of the agreement signed September 28, 1934, by B. B. Anthony of the Standard-Vacuum Oil Company, J. K. Bousfield of the Asiatic Petroleum Company (South China) Ltd., J. C. Williams of The Texas Company (China) Ltd., and Au Pong-po, Kwangtung

intrinsic importance than for the pattern of relationships it reveals. Stanvac, Asiatic, and The Texas Company maintained a solidly united front and concentrated on one objective—selling kerosene—while American and British diplomats appeared chiefly concerned with protection of treaty rights. Each concentrated on his institutional role and the two groups cooperated closely when their interests coincided. When their interests did not coincide, the diplomats occupied themselves elsewhere, and the oil companies used a combination of pressure, *de facto* bribery, and adroit psychology to cope with the intricate complications placed in their path. This identical pattern soon emerged on a considerably larger scale as expanding Japanese nationalism created increasing problems for the oil companies in Manchuria and Japan and Walden and Parker fought to protect Stanvac's investment in those areas.

THE POLITICAL complications behind Stanvac's difficulties in Manchuria far exceeded those in Canton. On September 18, 1931, an explosion on the South Manchurian Railroad had sparked fighting between the Japanese Kwantung Army and troops of the Chinese garrison north of Mukden. Engineered by a handful of extremists in the Kwantung Army with tacit approval from sympathizers in the Tokyo supreme command, the incident rapidly resulted in the full-scale occupation of Manchuria, while civilian authorities in Tokyo simultaneously attempted to restrain the Army and defend its actions to the outside world. International indignation mounted with the scale of fighting, the League of Nations repeatedly called for an end to the conflict, and Secretary of State Henry L. Stimson announced on January 7, 1932 that the United States would refuse to recognize any potential agreement between Japan and China that impaired American treaty rights or any change in the legal status of Manchuria brought about by the use of

---

Commissioner of Finance, FO371(1934)F6895/75/10. The details of the settlement were also reported to the Foreign Office in a letter from F. W. Starling (Director of the Petroleum Department of the Board of Trade) on October 31, 1934, FO371(1934)F6509/75/10.

force. Neither the League nor this "non-recognition" doctrine checked the Kwantung Army; a new state of Manchukuo was created in March of 1932 and recognized by Japan in September. That same fall the League's Lytton Commission returned from Manchuria with a report criticizing Japanese actions, and formal League approval of this report led to Japanese withdrawal from that body in February of 1933. A new state had been created, nominally independent, actually controlled by Japan, and completely unrecognized by the Western powers.[14]

Such was the political climate in the summer of 1933 when rumors began to circulate in Tokyo that the Japanese Army planned to press the Manchukuo government into establishing monopoly control over the production, refining, and distribution of petroleum products throughout Manchuria. Both Socony-Vacuum and the Department of State received remarkably detailed reports on these plans from their representatives in Japan but agreed that action would be premature until a monopoly materialized.[15] When rumor became reality in 1934,

[14]Among the many studies of the Manchurian Incident, three are particularly useful. Sadako N. Ogata, *Defiance in Manchuria: The Making of Japanese Foreign Policy, 1931–1932* (Berkeley: University of California Press, 1964) emphasizes the role of junior officers within the Army, while James Crowley in *Japan's Quest for Autonomy* suggests that higher military and civilian officials were not unhappy about the trend of events even though they did not instigate them. For the American reaction, see Armin Rappaport, *Henry L. Stimson and Japan, 1931–33* (Chicago: University of Chicago Press, 1963).

[15]Letter, Joseph C. Grew (American Ambassador in Japan) to SecState, May 12, 1933, RG 59, 894.6363/38; letter, M. S. Myers (American Consul General at Mukden) to SecState, July 10, 1933, *FRUS(1933)*, III, 736–37; letter, Kersey F. Coe (Socony-Vacuum Export Department) to Hornbeck, July 19, 1933, RG 59, 894.6363/47; letter, Grew to SecState, July 24, 1933, *FRUS(1933)*, III, 738–39; letter, Coe to Hornbeck, July 27, 1933, RG 59, 894.6363/46; letter, Coe to Hornbeck, July 31, 1933, RG 59, 894.6363/48; letter, Coe to Hornbeck, November 22, 1933, RG 59, 893.6363 Manchuria/9; letter, John Carter Vincent (American Consul at Dairen) to Grew, December 11, 1933, *FRUS(1933)*, III, 742–44. John C. Goold, Socony's manager in Yokohama, went so far as to send a representative to query personally the Japanese Vice Minister of War, Lt. Gen. Yanagawa, on the Army's plans.

the issue became rapidly entangled with parallel developments in Japan, where the Petroleum Industry Law of 1934 established extensive controls—but no monopoly. As a result, most published accounts, including documents in the *Foreign Relations* series, tended to equate the two issues or even to place greater emphasis on Manchuria.[16] A careful examination of the record, however, reveals a sharp difference in the reaction of diplomats and business interests to these two developments. The Department of State and Foreign Office did, indeed, demonstrate greater zeal for tackling the Manchurian problem, where the principle of the Open Door was at stake, but Stanvac and Asiatic devoted themselves almost exclusively to protection of their business in Japan. This is hardly surprising when one examines the figures in Tables II-1 and II-2 and finds that Stanvac's trade and investment in Japan amounted to sixteen times its interest in Manchuria.[17]

This disparity appears to have escaped the attention of diplomatic personnel, however, and their concern turned chiefly on violations of international law.[18] How the two issues

[16]In his memoirs, Cordell Hull struck out at the Manchurian monopoly but failed to mention parallel development in Japan; *The Memoirs of Cordell Hull* (2 vols.; New York: Macmillan, 1948), I, 275–76; while Joseph C. Grew at least gave equal attention to the two problems in his account, *Turbulent Era: A Diplomatic Record of Forty Years, 1904–1945*, edited by Walter Johnson (2 vols.; Boston: Houghton Mifflin, 1952), II, 942–43, 956–57, 967–70, 976–80, 1023–26. The same pattern appears in contemporary studies such as the Council on Foreign Relations, *The United States in World Affairs, 1934–1935*, by Whitney H. Shepardson (New York: Harper & Brothers, 1935), pp. 156–58; Claude A. Buss, *War and Diplomacy in Eastern Asia* (New York: Macmillan, 1941), p. 164; Elizabeth B. Schumpeter, ed., *The Industrialization of Japan and Manchukuo, 1930–1940: Population, Raw Materials and Industry* (New York: Macmillan, 1940), pp. 391–92, 433; and Irving S. Friedman, *British Relations with China, 1931–1939* (New York: Institute of Pacific Relations, 1940), pp. 47–49.

[17]More detailed statistics on the Manchurian trade are given in Appendix B, Table B-5.

[18]The zeal of the State Department in defending the Open Door in Manchuria despite the minimal commercial interests involved was noted by Charles C. Tansill in *Back Door to War: The Roosevelt Foreign Policy, 1933–1941* (Chicago: Henry Regnery, 1952), pp.

TABLE II — 1

COMPARATIVE VOLUME OF PETROLEUM SALES
BY ANGLO-AMERICAN COMPANIES IN
MANCHURIA AND JAPAN, 1933
(In thousands of 42-gallon barrels)

	Manchuria	Japan[a]
Asiatic Petroleum	253	5,960
Stanvac	197	3,128[b]
Texas Company	82	53
Total	532	9,087

[a]Including Korea and Formosa.

[b]Of this total, approximately half was sold to industrial and military customers through Mitsui Bussan Kaisha and the remainder through Stanvac's own distribution network.

SOURCE: Manchuria — Report of the British Consul General in Mukden to the British Legation in Peking, June 14, 1934; FO371(1934)F5158/142/10. Japan — Memorandum of August 20, 1934, provided to the Department of State by W. C. Teagle; RG 59, 894.6363/84.

became intertwined is a fascinating story, but relationships can be seen more clearly if the two are first treated separately. What follows, then, is an account of the Manchurian monopoly, with consideration of parallel developments in Japan deferred to the next chapter.

The Manchurian oil monopoly actually emerged in several stages, creating a situation almost as intricate and elusive as the one Stanvac had faced in Canton. On February 21, 1934, the Manchukuo government promulgated a law creating the

141–43, but a far more balanced account of the two episodes appears in Mira Wilkins, "The Role of U.S. Business" in Borg and Okamoto, eds., *Pearl Harbor as History*, pp. 341–76. Another, well-researched treatment of the problem appears in Jamie W. Moore, "Economic Interests and American-Japanese Relations: The Petroleum Monopoly Controversy," *The Historian*, XXXV (August 1973), pp. 551–67.

TABLE II — 2

COMPARATIVE VALUE OF THE STANDARD-VACUUM
OIL COMPANY'S INVESTMENT IN
MANCHURIA AND JAPAN, 1934
(Undepreciated first cost of fixed assets)

Manchuria	Japan[a]
$448,096	$7,108,587

[a]Including Korea and Formosa.
SOURCE: Manchuria — Ballantine (then American Consul
General at Mukden) to Johnson, May 11, 1935, reporting
claims submitted to the Manchukuo government, RG 59,
893.6363 Manchuria/215. Japan — Memorandum of
August 20, 1934, provided to the Department of State by
W. C. Teagle, RG 59; 894.6363/84.

Manchuria Petroleum Company for "producing, refining, and
selling" petroleum products throughout Manchuria, but with
no monopoly provisions. Headquarters were to be in Hsinking,
every aspect of the company's business could be government
controlled, and the initial capitalization of 5,000,000 Yuan
was to come twenty percent from the Manchukuo government,
forty percent from the South Manchurian Railroad, and ten
percent each from Mitsui, Mitsubishi, the Ogura Oil Company,
and the Nippon Oil Company. Shares could not be transferred
without approval of the board, a device designed to preclude
unwanted foreign participation. This company's major effort
proved to be the construction of a refinery and tank farm at
Kanchingtsu, opposite Dairen in the Japanese leased territory
of Kwangtung, an installation that finally began operations in
March 1935.[19]

[19]For translations of the Manchuria Petroleum Company law, see
letter, Myers to Johnson, March 6, 1934, RG 59, 893.6363 Manchuria
/13; and letter, P. D. Butler (British Consul General at Mukden) to
Cadogan, April 18, 1934, FO371(1934)F2689/142/10. The com-
mencement of Manchurian refinery operations was reported in a tele-

The Manchuria Petroleum Company itself was not a monopoly and provided no solid basis for protest, but diplomats and oil men reacted differently in this early phase. Until mid-July 1934, Stanvac management had the impression that the company would be able to sell in Manchuria under a quota system and appeared reconciled to competition from the new refinery. However, rumors that it would be forced out began to reach New York in July and by August 8 became firm enough for Arthur G. May of Stanvac's New York office to appeal

gram from Asiatic Petroleum's Shanghai office to its London office on March 27, 1935 (FO371(1935)F2085/94/23). It should be noted that a rumor went through Anglo-American diplomatic circles that the moving force behind the Manchurian monopoly was Hashimoto Keizaburo, president of the Nippon Oil Company and the new Manchuria Petroleum Company, and brother-in-law of General Hishikari, who was concurrently Commander of the Kwantung Army, Governor of the Kwantung Leased Territory, and Japanese Ambassador to Manchukuo (letter, Butler to Foreign Office, July 18, 1934, FO371 (1934)F4818/142/10; telegram, Johnson to SecState, August 28, 1934, *FRUS(1934)*, III, 725–26; and letter, Grew to SecState, October 17, 1934, *FRUS(1934)*, III, 741–42). An inference that Hashimoto was behind the monopoly is also contained in Francis C. Jones, *Manchuria Since 1931* (New York: Oxford University Press, 1949), p. 195, citing a *New York Times* story of October 25, 1934. A majority of reports in British and American archives, however, place prime responsibiity on the Kwantung Army rather than on Hashimoto, and Army influence appears more plausible. Ballantine, who had been transferred from Canton to the American Consulate General at Mukden, became convinced that the Army was the moving force, and so reported in February of 1935 (letter, Ballantine to SecState, February 23, 1935, RG 59, 893.6363 Manchuria/149). Grew appears to have been similarly convinced of military influence (memorandum of conversation with Clive, March 11, 1935, Volume 1, "Conversations, 1932–36", pp. 163–64, Joseph C. Grew MSS; Houghton Library, Harvard University, hereafter cited as "Grew MSS"; and letter, Grew to SecState, March 22, 1935, RG 59, 893.6363 Manchuria/163). The issue, of course, can not be resolved without research into appropriate documents of the Manchurian government and Kwangtung Army, if these can be located, and such an endeavor appears beyond the scope of this study.

to Hornbeck for State Department assistance.[20] Not only was assistance forthcoming, but Stanvac discovered that the British Foreign Office had already begun action *three months earlier.*[21] On April 26 the British Consul General at Mukden had been told by an official of the Manchukuo Department of Industry that "some form of monopoly" would "eventually" be adopted for petroleum products. This information triggered an urgent dispatch from Sir Alexander Cadogan, British Minister in China, to the Foreign Office, which promptly consulted the Board of Trade, the semi-independent Petroleum Department, and its Embassy in Tokyo, regarding a course of action.[22] British diplomats in both Peiping and Tokyo suggested concurrent action to their American counterparts, who checked with Washington[23] and reached agreement that this was "an

[20]Letter, Kersey F. Coe (Stanvac, New York) to Hornbeck, July 13, 1934, RG 59, 893.6363 Manchuria/22; letter, A. G. May (Stanvac, New York) to Hornbeck, August 8, 1934, RG 59, 893.6363 Manchuria/31. Asiatic Petroleum also originally thought it might be able to operate under a quota system (telegram, Cadogan to Foreign Office, June 13, 1934, FO371(1934)F3602/142/10).

[21]In a recent study of British interwar policy, William R. Louis concludes that the Foreign Office generally preferred to leave Japan alone in Manchuria because of the relative insignificance of British interests there, but he notes that the Foreign Office took a different line in the case of the oil monopoly (see his *British Strategy in the Far East, 1919–1939* [London: Oxford University Press, 1971], p. 224, n. 38).

[22]Letter, Butler to Cadogan, April 28, 1934, FO371(1934)F2820/142/10; telegram, Cadogan to Foreign Office, May 1, 1934, FO371(1934)F2476/142/10; telegram, Dodd (British Embassy, Tokyo), to Foreign Office, June 3, 1934, FO371(1934)F3324/142/10; letter, Starling to Foreign Office, June 4, 1934, FO371(1934)F3367/142/10; and letter (signature illegible) Commercial Relations and Treaties Department, Board of Trade to Foreign Office, June 7, 1934, FO371(1934)F3419/142/10.

[23]Telegram, Johnson to SecState, June 19, 1934, *FRUS(1934),* III, 710; telegram, Johnson to SecState, July 2, 1934, *FRUS(1934),* III, 711–12; telegram, Clive to Foreign Office, July 3, 1934, FO371(1934)F4025/142/10.

important test case of the principle of the Open Door in Man-chukuo,"[24] although care should be taken for "preservation of the policy of nonrecognition."[25] Accordingly, the British Am-bassador, Sir Robert Clive, and his American counterpart, Joseph C. Grew, presented notes of protest to the Japanese Foreign Office on July 2 and July 7 respectively, invoking the principle of the Open Door, Article III of the Nine-Power Treaty of 1922, Article XV of the Sino-American Treaty of 1844, and Article XIV of the Sino-French Treaty of 1858—each prohibiting the formation of monopolies in China that excluded foreign interests.[26] All this occurred *before* May's ap-peal to Hornbeck on August 8, and clearly explained subse-quent State Department readiness to support Stanvac on questions of treaty violations in Manchuria.

WHAT HAPPENED next provided a preview of events still seven years distant. Out of the Manchurian problem emerged the idea that Britain and America cooperate in an oil embargo to curb Japanese intransigence. While the urge toward such action temporarily subsided in 1935, the issues were clearly under-stood within the restricted circle of British and American oil-men and diplomats dealing with Manchuria and Japan in 1934, and the possibility of such action provoking Japanese seizure of the Netherlands East Indies was already clearly foreseen. An

[24]Entry of July 7, 1934, Joseph C. Grew Diary, Grew MSS.

[25]Memorandum of August 3, 1934, by E. H. Dooman in the Divi-sion of Far Eastern Affairs, RG 59, 893.6363 Manchuria/27; see also Foreign Office minutes of July 9, 1934, FO371(1934)F4125/142/10.

[26]Telegram, Clive to Foreign Office, July 2, 1934, FO371(1934) F3991/142/10; letter, Clive to Foreign Office, July 9, 1934, FO371 (1934)F4734/142/10; telegram, Grew to SecState, July 3, 1934, *FRUS(1934)*, III, 731–14; telegram, Hull to Grew, July 5, 1934, *FRUS(1934)*, III, 714–15; informal memorandum to the Japanese Ministry for Foreign Affairs, dated July 7, 1934, *FRUS(Japan, 1931–41)*, I, 130–31. For relevant text of the Nine-Power Treaty, see *FRUS (1922)*, I, 276; for the Sino-American Treaty of 1844, see Miller, *Treaties*, IV, 564; for the Sino-French Treaty of 1858, see Great Britain, Foreign Office, *British and Foreign State Papers* (London, 1812/14–), LI, 641.

56

increasing number of American and British officials joined in the debate, and considering the subsequent importance of the issue, it might be well to clarify the origins of the proposal.

The idea of using an embargo or other economic sanctions for the purpose of coercing Japan was not new, and had been widely debated in the Western press and in government circles at the time of the original Manchurian crisis in 1931.[27] Even the specific idea of an Anglo-American oil embargo to moderate Japanese treatment of American interests in Manchuria and Japan was not new in the fall of 1934. Rumors, or at least concern, over the possibility of such action had circulated in Tokyo for some time, and had prompted a candid discussion between First Secretary Dickover of the American Embassy in Japan, and Kurusu Saburo, Chief of the Commercial Affairs Bureau of the Japanese Foreign Office, as early as July 9, 1934. Kurusu asked Dickover about "a report . . . that the foreign oil companies were considering the placing of an embargo on the exportation of crude oil to Japan," and Dickover discounted such a possibility because "there were so many independent oil companies . . . that an oil embargo would hardly be practical." Kurusu noted that the independents could be controlled if "the oil companies . . . [could] . . . induce their governments to prohibit the exportation of crude to Japan and Manchuria" and asked if any such attempt were underway. When Dickover replied that he knew of none at the time but that such a move was always a "possibility," Kurusu pointed out that "the question of an embargo . . . was more serious than appeared at first glance, because it touched upon the question of national defense." The conversation then moved on to practical suggestions for resolving the Manchurian problem with minimum friction.[28]

[27]See, for example, Rappaport, *Stimson and Japan*, pp. 87–92.

[28]Memorandum of conversation, Dickover and Kurusu, July 9, 1934, *FRUS(1934)*, III, 715–18. Kurusu was an experienced professional diplomat who held a variety of key posts and was selected to join Ambassador Nomura for the final negotiations with the United States in Washington in the fall of 1941 (Hull, *Memoirs,* I, 276).

While formal consideration of an oil embargo may not have been underway in July, official Japanese repudiation of responsibility for events in Manchuria prompted serious Anglo-American consideration of just such a move in August. Almost identical notes from the Japanese Foreign Office were delivered to the American and British Embassies on August 2, asserting that Japan could not prevent her nationals from investing in the Manchuria Petroleum Company or dissuade the Manchukuo government from controlling the petroleum industry. Furthermore, any questions regarding the applicability of the treaties of 1844 and 1858 should be taken up directly with the government of Manchukuo[29]—which, of course, had not been recognized by either the United States or Britain. Some new form of leverage would have to be found if Stanvac and Asiatic were not to be forced out of Manchuria.[30] The Texas Company cooperated closely with the other two companies for a long period in Manchuria, but was not affected by developments in Japan, and does not appear to have actively engaged in consultations with the Department of State or the Foreign Office on either issue. In this instance the next initiative came from the Japanese offices of Stanvac and Asiatic.

Stanvac's resident manager in Tokyo, J. C. Goold, and his Rising Sun counterpart, H. W. Malcolm, held a series of conferences with Ambassadors Grew and Clive and concluded that they would be unable to make headway unless they could "use as [a] bargaining weapon [a] suggestion that foreign oil interests may be driven to retaliate by restricting supplies of . . . crude oil upon which [the] Manchurian refining programme and consequently the whole monopoly depends." They recognized that this might require "administrative action by [the] United States

[29]Letter, Clive to Foreign Office, August 8, 1934, FO371(1934) F5299/142/10; and telegram, Grew to SecState, August 3, 1934, *FRUS(1934)*, III, 719–20. The full text of the note to the American Embassy is given in *FRUS (Japan, 1931–41)*, I, 132–33.

[30]The British Foreign Office was in a real quandary as to how to proceed at this point (see extensive minutes attached to Clive's telegraphic report of August 8, 1934, FO371(1934)F4842/142/10).

Government to preclude [the] Japanese from resorting to alternative sources of supply in the United States" but considered that the attitude of the Japanese might be modified simply by the "knowledge that such a measure was under consideration."[31] This threat could take the subtle form of an innocuous "request by the American government for statistical data from American crude oil exporters for the purpose of studying the effect of crude oil exports on American oil interests abroad."[32] In view of the reverberations when an amplified version of this idea ran through Washington and London, it should be reemphasized that all Goold and Malcolm suggested was the *hint* of an embargo on crude to Japan and Manchuria as a psychological bargaining tool in *Manchuria*. The proposal for a *real* embargo that ultimately reached the British Cabinet was actually developed in the Foreign Office, and went well beyond anything proposed by the oil companies.

Grew and Clive forwarded the original suggestion to Washington and London with recommendations that it be given serious consideration. Grew indicated his endorsement in a strongly worded cable. He was "convinced that practical steps should be taken by the American Government and oil companies (working in conjunction with the British Government and oil companies) in an effort to defeat the proposal to establish an oil monopoly in Manchuria . . . [because] . . . a very serious issue is at stake, involving in large degree the future of our commercial interests and our traditional policy in the Far East."[33] The State Department appears to have been lukewarm or preoccupied, but its London counterpart reacted more

[31]The quotations are from Clive's telegraphic report of August 17, 1934, FO371(1934)F5068/142/10, but virtually the same message is contained in Grew's telegram to SecState of August 20, 1934, *FRUS (1934)*, III, 721–23. Confirmation that the idea of an embargo originated in a discussion of the Manchurian problem with Goold, Malcolm, and Clive is contained in Grew's diary entries for August 16 and 20, 1934, Grew MSS. See also Grew, *Turbulent Era*, II, 969.

[32]Telegram, Grew to SecState, August 20, 1934, *FRUS(1934)*, III, 721–23.

[33]Ibid.

warmly. Clive's comparatively modest cable to the Foreign Office sparked a full review within the British government on the feasibility and political desirability of a full embargo on crude directed against Japanese intransigence in both Manchuria *and* Japan.[34] The two problems had become inseparably intertwined, and with apologies for leaving the reader in suspense, further discussion of the embargo question will now be deferred to the next chapter. At this point it is significant to remember that the idea originated in the Tokyo offices of Stanvac and Rising Sun as a modest proposal for dealing with the Manchurian problem and received the initial endorsement of Grew and Clive as a means for defending the principle of the Open Door.[35]

THE DEVELOPMENT feared by Stanvac came one step closer on November 14, 1934, when the Manchukuo government promulgated a Petroleum Monopoly Law, leaving the date of enforcement up to the Minister of Finance, but providing legal

[34]Telegram, Clive to Foreign Office, August 17, 1934, with extensive Foreign Office minutes attached, FO371(1934)F5068/142/10. The risks of such action were not overlooked, even at the outset. The original Foreign Office minutes of August 19 by Sir Pratt note that, "The really serious aspect of the matter is . . . the possible effect upon the minds of the military of realizing that their supplies of an essential munition of war might at any moment be cut off. They might decide that considerations of national security justified them in forcibly seizing the . . . oil wells of the Netherlands East Indies. This seems a little far-fetched . . . [but] . . . there has already been considerable nervousness in the Netherlands East Indies as to Japanese designs." In balance, however, Pratt concluded that "the risks involved . . . are not so great or so certain as to cause us to reject a proposal which seems to provide us with our only weapon for defending our intrests in a vast and rapidly expanding market." As will be seen in the next chapter, the Admiralty considered the risk far more serious than did the Foreign Office.

[35]Although no formal content analysis has been made, this exact phrase, "the principle of the Open Door," recurs frequently enough in Anglo-American diplomatic correspondence to suggest a value almost as completely internalized as "the right of self-defense." A study of this hypothesis could prove interesting.

machinery for establishing a Petroleum Monopoly Bureau, restricting sales in Manchuria to designated agents or licensees, and providing for the purchase of stocks and equipment of existing dealers if application were made within one month after implementation.[36] The law made no official mention of the Manchuria Petroleum Company or methods for selecting "designated sales agents," but private inquiries revealed the real intent. The plan called for absorbing the full output of the new Dairen refinery and the Fushun shale oil plant plus importation of additional refined products to meet total requirements. All existing bulk distribution facilities in Manchuria would be taken over by the Monopoly Bureau except those owned by Stanvac, Asiatic, and Texas at Newchwang and Dairen, which were not needed. Sales throughout Manchuria, including the South Manchurian Railroad Zone, would be handled by ten principal agents in selected cities and numerous subagents picked from among present distributors. Agents of foreign companies could participate at this level, but only by selling monopoly stocks under monopoly brand names at monopoly prices. Stanvac, Asiatic, and Texas would thus lose control of their long-standing distribution networks but were offered preferential treatment in the purchase of crude for the new refinery and whatever additional refined products might be needed.[37] In short, the stakes included a reduction in their profits (the margin on crude was lower than that on refined),

---

[36]For translations of this and two supplementary laws, see letters, Butler to Sir Robert Clive (British Ambassador in Japan), November 15, 1934, and November 16, 1934, FO371(1934)F7171/1659/23, and FO371(1934)F7173/1659/23.

[37]Letter, L. B. Loucks (Stanvac Mukden manager) to A. S. Chase (American Consulate General, Mukden), October 30, 1934, RG 59, 893.6363 Manchuria/98; letter, T. G. Ely (Rising Sun Petroleum Company, Yokohama) to H. A. MacRae (British Embassy, Tokyo), November 9, 1934, British Embassy and Consular Archives: Japan, Class F.O.262, Public Record Office, London, hereafter cited as "FO262," folder FO262/1883, "Oil: Manchuria, 1934," pp. 21–39; letter, Butler to Clive, November 29, 1934, FO371(1934)F7468/1659/23.

the loss of assured outlets through their own sales networks, and violation of the principle that they should not be forced out of the distribution business by a government monopoly.

Meanwhile, the Manchurian Oil Company had asked for bids on crude for its Dairen refinery, Stanvac, Rising Sun, and Texas had refused to quote, and the void had immediately been filled by bids from Standard Oil of California and the Union Oil Company on Kettleman crude—exactly the quality desired.[38] For six months, while the Dairen refinery was being completed, Manchurian and Japanese officials repeatedly tried to reach some sort of business compromise with Stanvac, Asiatic, and Texas—even offering a guaranteed annual quota to be purchased by the monopoly. The best offer involved a guarantee by the Monopoly Bureau to purchase 167 thousand barrels annually from the three oil companies (about one-third of their 1933 sales), conversion of this into a fixed percentage of the trade within two or three years, and preference to the companies in purchases of crude for the new refinery. But Walden and Parker and their counterparts in the other two companies stood firm—if they could not retain control of their own distribution systems in Manchuria, they would have no dealings with any Manchurian agency.[39]

Barring an embargo backed by the United States govern-

[38]Telegram, Grew to SecState, August 31, 1934, *FRUS(1934)*, III, 727–28; letter, Butler to Cadogan, September 3, 1934, FO371(1934) F5677/142/10; telegram, Grew to SecState, September 5, 1934, *FRUS(1934)*, III, 729; letter, Butler to Cadogan, October 6, 1934, FO371(1934)F6270/142/10; and letter, Grew to SecState, November 30, 1934, RG 59, 893.6363 Manchuria/113.

[39]Telegram, Gauss (American Chargé in China) to SecState, October 21, 1934, *FRUS(1934)*, III, 742–43; letter, Butler to Cadogan, October 23, 1934, FO371(1934)F6664/1659/23; telegram, Gauss to SecState, November 1, 1934, *FRUS(1934)*, III, 753; telegram, Clive to Foreign Office, November 29, 1934, FO371(1934)F7076/1659/23; telegram, Gauss to SecState, November 29, 1934, *FRUS(1934)*, III, 771–72; Kersey F. Coe (Stanvac, New York) to Hornbeck, December 7, 1934, RG 59, 893.6363 Manchuria/104; telegram, Grew to SecState, December 13, 1934, *FRUS(1934)*, III, 787; letter, Grew to SecState, December 13, 1934, *FRUS(1934)*, III, 787–88; J. B. Loucks (Stanvac, Mukden)

ment, the only counter weapon available was words, and these were used abundantly by British and American diplomats. Grew and Clive registered increasingly strong protests in Tokyo, only to be met with polite disclaimers of responsibility,[40] a tactic that infuriated American Secretary of State Cordell Hull. In his *Memoirs* Hull recalled that "Japan was flagrantly dishonest in her statements regarding Manchuria. She had conquered Manchuria, set up a puppet Government there with the strings manipulated by her army, and occupied it with her troops. Yet [Foreign Minister] Hirota had the audacity to say,

---

to Stanvac, Shanghai, December 13, 1934, RG 59, 893.6363 Manchuria /129; telegram, Clive to Foreign Office, February 7, 1935, FO371 (1935)F852/94/23; letter, Grew to SecState, February 7, 1935, RG 59, 893.6363 Manchuria/141; telegram, Grew to SecState, February 8, 1935, *FRUS(1935)*, III, 879–80; telegram, Grew to SecState, February 21, 1935, *FRUS(1935)*, III, 882; telegram, Phillips (Acting Secretary of State) to Grew, February 23, 1935, *FRUS(1935)*, III, 882; letter, H. Dundas (Stanvac, New York) to Hornbeck, February 25, 1935, RG 59, 893.6363 Manchuria/142; telegram, Grew to SecState, March 2, 1935, *FRUS(1935)*, III, 884; telegram, Clive to Foreign Office, March 2, 1935, FO371(1935)F1441/94/23; the most complete account of the final commercial negotiations in Tokyo between Walden and Parker (Stanvac), Godber (Shell), Kurusu (Japanese Foreign Office), Hoshino (Manchukuo Vice Minister of Finance), and Tsuge (Petroleum Advisor to the Manchukuo government) is in a report from Grew to SecState, March 6, 1935, RG 59, 893.6363 Manchuria /152, but see also letter, Clive to Foreign Office, March 4, 1934, FO371(1935)F2125/94/23 for a briefer version of the same information.

[40]Formal British protests were delivered to the Japanese Foreign Office on July 1, August 21, and November 24, 1934; letter, Clive to Foreign Office, July 9, 1934, FO371(1934)F4734/142/10; letter, Clive to Foreign Office, August 31, 1934, FO371(1934)F5877/142/10; and telegram, Clive to Foreign Office, November 24, 1934, FO371(1934) F6971/1659/23. For the Japanese replies, see letter, Clive to Foreign Office, August 8, 1934, FO371(1934)F5299/142/10; letter, Clive to Foreign Office, November 9, 1934, FO371(1934)F7160/1659/23; and letter, Clive to Foreign Office, March 26, 1935, FO371(1935)F2527/ 94/23. Parallel American notes were delivered on July 7, August 31, and November 30 with similar replies from the Japanese government; *FRUS(Japan, 1931–41)*, I, 130–48.

in a memorandum to us on November 5, 1934; 'The plan of the Government of Manchukuo for the control of the oil industry is a project of that Government itself and is not within the knowledge or concern of the Imperial [Japanese] Government, and the Imperial Government is not in a position to give any explanation with respect to it.' "[41] At least one American historian has commented that this "friction between the United States and Japan over Japanese commercial policies in Manchukuo was entirely needless,"[42] but considering how strongly American diplomats felt committed to the Open Door, it was probably inevitable.

Despite the rhetorical thunder, the end came quietly on April 10, 1935, when the monopoly went into effect. The three companies withdrew from Manchuria, and the Dairen refinery proceeded to procure supplies from independent producers in California. Simple as this sounds, documents in British and American archives suggest there was a low-keyed Machiavellian touch to the withdrawal, instigated by Parker. As soon as it became clear that the monopoly would actually go into effect, intensive consultations began in Tokyo, London, New York, and Washington on how and where to file damage claims. At the outset, Sir Andrew Agnew of Shell's London office made it clear that from his viewpoint "the position in the two countries should now be kept separate," and the companies should not "sacrifice the Japanese market merely for the sake of pressure on Manchukuo."[43] In other words, any action taken in Manchuria should be supportive of the companies' position in Japan, where the commercial stakes were far more critical. The issue debated by oil men and diplomats was how this could best be accomplished. Parker, the sophisticated veteran of twenty years in the China market, had an idea.

After consultation with Stanvac personnel from Shanghai

[41]Hull, *Memoirs,* I, 275.

[42]Tansill, p. 143.

[43]Foreign Office minutes of April 4, 1935, by C. W. Orde, recording a meeting with Agnew (Shell), Darch (Asiatic Petroleum), and Starling (Petroleum Department), FO371(1935)F2227/94/23.

and Mukden, Parker, who was then in Yokohama, cabled New York suggesting an offer to sell the company's Manchurian facilities to the Monopoly Bureau and presentation of a stiff damage claim against the Japanese government through diplomatic channels.[44] This suggestion precipitated a daylong conference between Stanvac and State Department representatives in Washington, with telephone calls to New York and London to ensure a consolidated position, and Parker was advised to try for a comprehensive settlement in Manchuria, exhausting that avenue before considering diplomatic pressure against Japan.[45] Parker cabled strong objections, but as an alternative proposed an offer to sell with a statement that the "value of properties could not be separated from value of whole business as a going concern which . . . will be stated in one lump sum," calculated by consolidating physical equipment at first cost, real estate at cost plus appreciation, and five or ten years of anticipated future earnings.[46] He wanted to file

[44]Parker's conference in Yokohama with Twogood (manager of Stanvac's North China Division) from Shanghai, and Jones and Bristow from Stanvac's Mukden office, was reported in a letter from Grew to SecState, May 15, 1935, RG 59, 893.6363 Manchuria/212; Parker's initial suggestion is contained in a cablegram to Dundas (Stanvac, New York), April 22, 1935, copy in RG 59, 893.6363 Manchuria/184. The full text of Parker's telegram was immediately passed on from New York to Asiatic Petroleum's London office; telegram, "H.M." (New York) to "NTL ASDEBATIC" (London), April 22, 1935, copy in FO371(1935)F2732/94/23.

[45]Memorandum of conversation, Hornbeck, Mackay, and Collins (Department of State) with Arthur G. May and Claude A. Thompson (Stanvac), April 24, 1935, RG 59, 893.6363 Manchuria/184. Confirmation of a consolidated position is contained in the Foreign Office minutes attached to a telegram, Clive to Foreign Office, April 26, 1935, FO371(1935)F2722/94/23. The instructions to Parker are quoted in a telegram, Hull to Grew, April 24, 1935, FRUS(1935), III, 906, and confirmed in a letter, Thompson to Hornbeck, April 25, 1935, RG 59, 893.6363 Manchuria/181.

[46]Cablegram, Parker to Dundas, April 26, 1935, copy in RG 59, 893.6363 Manchuria/137. See also the simpler version of Parker's counterproposal reported in a telegram from Grew to SecState, April 26, 1935, FRUS(1935), III, 907.

duplicate claims in Manchuria and Japan, but after further consultation in Washington, the New York office overruled this procedure on the theory that a stronger diplomatic protest could be made later if the American and British governments had not already "associated themselves with a compromise by filing with the Japanese government" an offer to sell "physical or other assets to Manchukuo."[47]

Further cables between New York, Washington, London, and Tokyo ensured a united front,[48] and on May 10, 1935, Stanvac, Asiatic, and The Texas Company submitted claims to the Manchukuo government in Hsinking for the "total value of their businesses as going concerns" with an inventory of property and equipment and no indication in the claim itself of how "total value" had been calculated. The amounts were as shown in Table II–3.[49]

Joseph W. Ballantine, recently transferred from Canton to the post of American Consul General at Mukden, reported that the companies had "no serious expectation that the 'Manchukuo' Government will recognize responsibility for the 'total

[47]The Washington discussions are covered in a memorandum of conversation, Hornbeck and Mackay (State Department) with Dundas, Thompson, and May (Stanvac), April 29, 1935, RG 59, 893.6363 Manchuria/216; the resulting instructions to Stanvac's Japanese office are quoted in a telegram, Hull to Grew, May 1, 1935, FRUS(1935), III, 909–10; and the quotation on strategy is taken from an April 30 cable to Asiatic Petroleum attached to a letter from May to Hornbeck, May 3, 1935, RG 59, 893.6363 Manchuria/198.

[48]Telegram, Hull to Grew, May 1, 1935, FRUS(1935), III, 910–11; telegram, Hull to Bingham (American Ambassador in London), May 2, 1935, FRUS(1935), III, 911–12; letter, Dundas to Hornbeck, May 2, 1935, RG 59, 893.6363 Manchuria/192; letter, May to Hornbeck, May 3, 1935, RG 59, 893.6363 Manchuria/198; telegram, Hull to Bingham, May 3, 1935, FRUS(1935), III, 912; telegram, Bingham to SecState, May 4, 1935, FRUS (1935), 913; and telegram, Hull to Grew, May 4, 1935, FRUS(1935), III, 913.

[49]Letter, Ballantine to Johnson, May 11, 1935, RG 59, 893.6363 Manchuria/215; see also letter, Grew to SecState, May 15, 1935, RG 59, 893.6363 Manchuria/212; telegram, Butler to Foreign Office, May 10, 1935, FO371(1935)F3004/94/23; and letter, Butler to Clive, May 11, 1935, FO371(1935)F4336/94/23.

TABLE II — 3

STANVAC, TEXAS AND ASIATIC
CLAIMS SUBMITTED TO THE MANCHUKUO GOVERNMENT
MAY 10, 1935

Company	Property and Equipment	Total Value
Stanvac	$448,096	$1,781,881
Texas	$ 40,645	$ 456,738
Asiatic	£145,687	£ 418,000

SOURCE: Letter, Ballantine to Johnson, May 11, 1935, RG 59, 893.6363 Manchuria/215.

value' claimed by them, and are presenting their claims primarily in order to exhaust local remedies before requesting diplomatic action by their governments."[50]

As might be expected, the Manchukuo government displayed no eagerness to come to terms on this basis, and on November 13 representatives of the three oil companies called at the Manchukuo Ministry of Finance in an attempt to obtain an unequivocal response.[51] When this failed, the American and British Consuls General in Mukden called at the Manchukuo Ministry of Foreign Affairs on December 4 for the same purpose.[52] Finally, on December 13, in a letter to Ballantine, the Vice Minister of Foreign Affairs put his government's position in writing: While Manchukuo would "be prepared to purchase the . . . equipment for a reasonable price," it would not "recognize the principle of compensating the losses in-

[50]Letter, Ballantine to Johnson, May 11, 1935, RG 59, 893.6363 Manchuria/215.

[51]Memorandum of conversation at the Ministry of Finance, Hsinking, Manchuria, November 13, 1935, Bristow (Stanvac), Mott (Rising Sun), and Menefee (Texas Company) with Hoshino and Namba (Ministry of Finance), attached to a letter from Bristow (in Dairen) to Stanvac, Yokohama, November 14, 1935, copy in RG 59, 893.6363 Manchuria/266.

[52]Telegram, Johnson to SecState, December 6, 1935, *FRUS(1935)*, III, 933; and letter, Butler to Clive, December 5, 1935, FO371(1935) F8068/94/23.

curred by your oil companies" over and above the actual value of facilities.[53] When this reply reached New York and Washington, Parker wrote Hornbeck suggesting that "all local recourse has been exhausted," and "the time has now arrived for a claim to be placed with the Japanese government."[54] Hornbeck agreed, and Stanvac's legal department began preparation of a formal diplomatic claim, to be held in abeyance pending a decision "in regard to the oil situation in Japan."[55]

Whether planned or not, the withdrawal from Manchuria had been converted into a diplomatic claim against Japan which could be used at the company's discretion with assurance that the State Department would press the matter since it involved a violation of treaty rights. Parker permitted the issue to lie dormant for two years, and then, as will be seen in the next chapter, resurrected it to bring "pressure upon the Japanese government . . . [to] . . . give the company additional leverage in connection with its negotiations relating to business in Japan."[56] While the available documents do not permit

[53]The text of this letter is quoted in a telegram, Johnson to SecState December 16, 1935, *FRUS(1935)*, III, 936–37; and a copy is attached to a letter, Ballantine to Johnson, December 16, 1935, RG 59, 893.6363 Manchuria/277.

[54]Letter, Parker (in New York) to Hornbeck, December 24, 1935, RG 59, 893.6363 Manchuria/274; see also Grew's telegram to SecState, December 21, 1935, *FRUS(1935)*, III, 938; and telegram, Clive to Foreign Office, December 20, 1935, FO371(1935)F7958/94/23.

[55]Letter, Hornbeck to Parker, December 30, 1935, RG 59, 893.6363 Manchuria/274; telegram, Hull to Grew, December 30, 1935, *FRUS (1935)*, III, 939; memorandum of conversation, Hornbeck, Hamilton, and Mackay (State Department) with Thompson (Stanvac), January 9, 1936, RG 59, 893.6363 Manchuria/279; and memorandum of conversation, Mackay with Thompson, January 24, 1936, *FRUS(1936)*, IV, 787–88 (the source from which the quotation is taken).

[56]Memorandum of conversation, Hornbeck, Dooman, Vincent, and Ballantine (State Department) with Parker and Marshall (Stanvac), April 15, 1937, RG 59, 893.6363 Manchuria/294. As matters turned out, Stanvac's Manchurian claim surfaced periodically but remained unsettled when hostilities commenced in 1941; The Texas Company

absolute certainty, circumstantial evidence strongly suggests that this had been Parker's strategy from the very first. Despite diplomatic rhetoric, the oil companies clearly considered their problems in Japan far more important than those in Manchuria, and Stanvac appears to have handled the Manchurian withdrawal chiefly with an eye toward the leverage it could provide in Japan.[57]

To RECAPITULATE, an embryonic pattern of cooperation between Stanvac, Shell, the Department of State, and the Foreign Office had emerged in East Asia at least as early as 1933, with The Texas Company as a silent partner on matters relating to China. The Canton kerosene war demonstrated the dedication of oil men and diplomats to their respective institutional roles —whether it was selling kerosene or defending treaty rights. When the two roles intersected, cooperation proved impecable, but when the problem moved outside the scope of international law, the oil companies used whatever leverage they could muster to keep the market open. By 1934 and 1935 a similar pattern had begun to show itself in Manchuria, when plans to

---

made a partial settlement by disposing of its physical properties sometime prior to August 1937 (telegram, Grew to SecState, August 21, 1937, *FRUS(1937)*, IV, 734); and Asiatic's claim was recorded as still unsettled as late as May 1940 (letter, F. W. Starling [Petroleum Department] to Ashley Clarke [Foreign Office] May 22, 1940, FO371 (1940)F3303/1073/10). The united front of the three companies against sales to the Monopoly Bureau held until May of 1936, when The Texas Company defected (letter, Grummon [American Consul at Dairen] to Grew, April 13, 1936, RG 59, 894C.6363/23). What Stanvac and Asiatic did after that point is unclear from the available records.

[57]This interpretation is slightly different from that in Mira Wilkins' previously mentioned article on "The Role of U.S. Business," in Borg and Okamoto, *Pearl Harbor as History,* p. 365. Dr. Wilkins' conclusions on the Manchurian withdrawal are based on documents published in the *Foreign Relations* series, which imply that Stanvac and Shell made a real effort to sell. Information in the unpublished correspondence suggests otherwise, although the two companies probably would not have refused an offer to meet their rather high price.

establish an oil monopoly provoked vigorous protests from the Foreign Office and State Department in defense of treaty rights and the principle of the Open Door. The oil companies, however, gave first priority to the far larger market in Japan and handled the Manchurian issue primarily with an eye to the effect it would have in Tokyo. Two specific ideas emerged from Manchuria and were promptly converted into weapons for the larger arena—the hint of an oil embargo and a diplomatic claim against Japan for "damage" inflicted by Manchukuo. To appreciate the full scope of the pattern that had begun to take shape, it is now necessary to turn to Japan and the events precipitated by the Petroleum Industry Law of 1934.

THREE

AND JAPAN'S "QUEST
FOR AUTONOMY"

BY FAR THE most serious East Asian problem faced by Walden and Parker in the early 1930s was the possibility of losing all or part of Stanvac's extensive Japanese market as a result of mounting pressure within Japan to decrease dependence on foreign refineries. The problem centered on the interpretation and implementation of the Japanese Petroleum Industry Law promulgated in March 1934. Understanding the extent and complexity of the forces with which Walden and Parker had to contend requires a brief examination of the mood of Japan in the early 1930s and the specific origins of the troublesome law. The petroleum law was only one of many symptoms of Japan's growing sense of isolation and its desire for as much autonomy as possible in what the Japanese perceived as an increasingly hostile world.

Some insight into the political climate of the period may be gained by examining the famed "Amau Doctrine" of 1934. On April 17, 1934, three weeks after passage of the Petroleum Industry Law, Amau Eiji, Chief of the Bureau of Information and Intelligence of the Japanese Foreign Office, casually read to a group of newsmen a document intended to clarify Japan's position on pending questions of third-power financial and military aid to China. The thrust of the statement had been openly discussed in Japanese government circles for some time, but the viewpoint generated no real reaction from the diplomatic corps until emphasized by Amau's press conference.[1] The phraseology, in effect, constituted a Japanese Monroe

[1]Crowley, *Japan's Quest for Autonomy*, p. 197; the title for this chapter is derived from Crowley's excellent study. For the American reaction to the "Amau Doctrine" see Borg, *The United States and the Far Eastern Crisis*, pp. 46–99.

Doctrine for East Asia and prompted a flurry of disturbed discussion in Washington and London:

> Owing to the special position of Japan in her relations with China, her views and attitude respecting matters that concern China, may not agree in every point with those of foreign nations: but it must be realized that Japan is called upon to exert the utmost effort in carrying out her mission and in fulfilling her responsibilities in East Asia.
>
> Japan has been compelled to withdraw from the League of Nations because of their failure to agree in their opinions on the fundamental principles of preserving peace in East Asia. Although Japan's attitude toward China may at times differ from that of foreign countries, such differences cannot be evaded, owing to Japan's position and mission.
>
> It goes without saying that Japan at all times is endeavoring to maintain and promote her friendly relations with foreign nations, but at the same time we consider it only natural that, to keep peace and order in East Asia, we must act alone and on our own responsibility. . . . We oppose therefore any attempt on the part of China to avail herself of the influence of any other country in order to resist Japan . . . .
>
> The foregoing attitude of Japan should be clear from the policies she has pursued in the past . . . [but] it is deemed not inappropriate to reiterate her policy at this time.[2]

The sense of special "mission" and determination to "act alone" had been building for some time, and ultimately found expression in the mystique of Japan's "New Order in East Asia" and the Imperial Rescript for War of December 8, 1941.[3] While the consequences of this mood were not clearly

[2]From the translation which Grew dispatched to the Department of State, *FRUS (Japan 1931–41)*, I, 224–25.

[3]The brief Imperial Rescript asserted that Japan's basic policy was "to ensure the stability of East Asia," and concluded that owing to

discernible in 1934, some of the reasons for it were: Japanese views of the outside world in the 1930s were a compound of resentment over Western racial attitudes, expressed in such actions as increasingly restrictive American immigration laws; economic discontent (especially agrarian) aggravated by worldwide depression; and a feeling that Japan had no real allies among the great powers—enhanced by the difficulty experienced in obtaining Allied recognition of its Shantung acquisitions at Versailles, the end of its 1902 alliance with Britain at the Washington Conference of 1921–22, and the refusal of the Western powers to permit an increase in Japan's allocation of heavy cruisers at the Washington Naval Conference of 1930.[4] When the Kwantung Army engineered the Manchurian Incident in 1931, Tokyo attempted to restrain it, but closed ranks to present a unified front to the outside world, and was dismayed by worldwide criticism when the Manchuria affair became a *fait accompli*. As James Crowley has put it: "Japan was now isolated, censured by the Western sea powers, hated by the Chinese, and opposed by the Soviet Union. Previously, whenever the powers had united against Japan's continental policy, Japan had always adjusted its diplomacy according to their dictates. This time, however, Japan was prepared to go it alone, to make its own destiny in Asia, independent of the wishes or guidance of the Occidental

---

the trend of events the "Empire for its existence and self-defense has no other recourse but to appeal to arms" (Archives of the Japanese Ministry of Foreign Affairs, 1868–1945; selected documents microfilmed for the U.S. Library of Congress, 1949–1951; microfilm reel WT 59, document IMT 451; Library of Congress, Washington, D.C.).

[4]Crowley argues that the political debate triggered by Premier Hamaguchi Yukō's acceptance of a 10:6 rather than a 10:7 ratio in cruisers far outweighed its actual military significance and severely weakened the position of those Japanese—chiefly civilians—who wanted to base Japan's security on cooperation with the West. Hamaguchi himself was shot and severely wounded by a young fanatic in November 1930, ushering in a decade of political violence (Crowley, pp. 78–81).

powers."[5] Increasingly, this mood linked Japan's destiny with its continental position, created a demand for military and economic autonomy, and fed the fires of ultranationalism.

Although there was no direct connection with the Amau Doctrine of April 1934, the same mood had prompted the Japanese Diet to enact a Petroleum Industry Law in March 1934, designed to buttress Japan's precarious position in strategic fuel. Antonomy in a hostile world required as much self-sufficiency as possible, and Japan was critically weak in petroleum reserves. Domestic production on a limited scale dated from 1880, but had actually declined after 1916, and by 1934 Japan had direct control over only a few low-producing fields in northwestern Honshu, Formosa, and Sakhalin, plus the Fushun shale oil plant in Manchuria, providing a combined total of about seven percent of her actual requirements.[6] The strategic implications of this deficit were obvious to all directly concerned. As early as 1933, Grew reported a private statement by an official in the Japanese Foreign Office

[5]James B. Crowley, "A New Deal for Japan and Asia: One Road to Pearl Harbor," in James B. Crowley, ed., *Modern East Asia: Essays in Interpretation* (New York: Harcourt, Brace, & World, 1970), p. 245. In addition to Crowley's longer work, *Japan's Quest for Autonomy*, four others are especially useful on Japan in the 1930s: Butow, *Tojo and the Coming of the War;* Hugh Byas, *Government by Assassination* (New York: Knopf, 1942); Yale C. Maxon, *Control of Japanese Foreign Policy: A Study of Civil-Military Rivalry, 1930–1945* (Berkeley: University of California Press, 1957); and Rōyama Masamichi, *Foreign Policy of Japan: 1914–1939* (Tokyo: Japanese Council, Institute of Pacific Relations, 1941). For an analysis of Japanese policy in the preceding decade, see Iriye, *After Imperialism.*

[6]Mitsubishi Economic Research Bureau, *Japanese Trade and Industry: Present and Future* (London: Macmillan, 1936), p. 212; Schumpeter, *The Industrialization of Japan and Manchukuo*, p. 774; letter, Grew to SecState, April 21, 1933, RG 59, 894.6363/32; and Table B-6, Appendix B. Standard Oil's International Petroleum Company had participated in Japanese production around the turn of the century, but in 1906 was bought out by Nippon Petroleum, which through mergers went on to become the dominant company in Japanese domestic production.

74

that "the Japanese Navy would seize the Dutch East Indies oil fields immediately on the outbreak of a war, no matter who the enemy might be."[7]

This deficit was a serious matter, but the original impetus for a control law came from an altogether different direction. Economic depression had produced chaos in the Japanese petroleum business, and a general movement was afoot in the early 1930s to rationalize key sectors of the economy for efficiency and price stability. An Important Industries Control Law encouraging cartelization had been enacted in April 1931 and administratively extended to encompass the petroleum business in November 1932. Nippon, Kokura, Mitsubishi, Mitsui, Rising Sun, and Socony-Vacuum negotiated a price and division of sales agreement, but the refusal of a marketer of Russian oil to cooperate continued to generate instability.[8] Meanwhile, international criticism of the Manchurian episode sent a shock wave through government circles, and in the summer of 1933 a joint committee for development of a national oil policy was established with representation from the Ministries of Navy, Army, Finance, Commerce and Industry, Foreign Affairs, and Colonial and Overseas Affairs. Business interests wanted regulation to assure price stability, and the Japanese government wanted regulation as a means for developing as autonomous a position as possible in petroleum refining and reserve stockpiling. The two movements coalesced on March 27, 1934, with promulgation of the Petroleum Industry Law, but military demands for a quasi-wartime economy

---

[7]Letter, Grew to SecState, April 21, 1933, RG 59, 894.6363/32; the official was identified as Mr. Shiratori, and the comment was reported as an illustration of ideas then being discussed rather than as an indication of firm plans.

[8]Schumpeter, *The Industrialization of Japan and Manchukuo*, pp. 774–75; George C. Allen, *Japanese Industry: Its Recent Development and Present Condition* (New York: Institute of Pacific Relations, 1940), p. 17; letter, Grew to SecState, April 21, 1933, RG 59, 894.6363/32; letter, F. S. Fales (president of Socony-Vacuum) to SecState, May 10, 1933, RG 59, 894.6363/35.

gradually replaced business pressure for price stability as the moving force behind its administration.[9]

As enacted, the Petroleum Industry Law established a licensing system for the production, refining, marketing, and importation of oil, empowered the government to set quotas, fix prices, and make compulsory purchases, and required all companies operating in Japan to maintain stockpiles of a quantity to be fixed by the government. Nothing in the law itself discriminated against or was clearly detrimental to foreign companies, and Stanvac, Rising Sun, and the American and British Embassies initially restricted themselves to worried inquiries as to how the law would be administered.[10] As administrative rules unfolded, however, it became clear that Stanvac and Rising Sun had real problems. In the original debate in the lower house of the Diet, one of the proponents had stated that "the primary objectives of the law were (1) to secure an adequate supply of oil at all times and (2) to regulate imports of oil with a view to promoting the development of the oil refining industry in Japan,"[11] and this explanation proved an accurate forecast of the manner in which the law was ex-

[9]Mitsubishi Economic Research Bureau, p. 216; Schumpeter, *The Industrialization of Japan and Manchukuo*, pp. 774–76; Allen, *Japanese Industry*, pp. 16–23. See also letter, Grew to SecState, September 6, 1934, RG 59, 894.6363/78, describing the original debate in the Diet. A major argument was the need to acquire as much autonomy as possible in oil reserves.

[10]Telegram, Grew to SecState, March 5, 1934, *FRUS(1934)*, III, 700–701; telegram, Grew to SecState, March 29, 1934, *FRUS(1934)*, III, 704–705; letter, Grew to SecState, April 5, 1934, RG 59, 894. 6363/61; and letter, Sir F. O. Lindley (British Ambassador in Tokyo) to Sir John Simon (Foreign Secretary), April 25, 1934, FO371(1934) F3153/1659/23. Translations of the Petroleum Industry Law are attached to both of the latter two documents. See also Grew, *Turbulent Era*, pp. 956–57.

[11]Letter, Grew to SecState, April 5, 1934, RG 59, 894.6363/61; see also telegram, Grew to SecState, March 29, 1934, *FRUS(1934)*, III, 704–5, and memorandum of conversation with the Japanese Vice Minister for Foreign Affairs, April 2, 1934, Volume 1, "Conversations, 1932–36," pp. 63–64, Grew MSS.

ecuted. An Imperial Ordinance of June 26 designated the Ministry of Commerce and Industry as executor, and rules issued as a Ministerial Ordinance on June 27 included a requirement that all companies operating in Japan build up and maintain at their own expense a stockpile equal to six months of normal sales volume.[12] Stanvac estimated that compliance would cost at least $375,000 for additional tankage plus $1,900,000 for stocks unnecessarily tied up; the cost to Rising Sun would be considerably more because of a higher sales volume and proportionately less tankage already in place.[13] To

[12]Enclosures in letter from Sir Robert Clive (the new British Ambassador in Tokyo) to Simon, November 9, 1934, FO371(1934) F6362/1659/23; this translation of the Ministerial Ordinance is marked June 17, but the date is given as June 27 elsewhere, and the latter date is more plausible. This manner of establishing the stock-holding requirement made the requirement technically more negotiable than a Law or Imperial Ordinance, but it represented the type of negotiated consensus within the Japanese government that outsiders found almost impossible to challenge.

[13]Memorandum of August 21, 1934, provided to the State Department by Walter C. Teagle, President, Standard Oil Company (New Jersey), RG 59, 894.6363/84.

[14]As analyzed by Royal Dutch-Shell, the impact of the new quotas for gasoline was as follows:

Percent of Market

	Actual Sales, 3 months through August 1934	Quotas for July through December 1934
Stanvac	19.18	17.81 (decrease)
Rising Sun	29.47	27.65 "
Matsukata (Russian Oil)	4.94	4.64 "
Nippon	21.87	23.93 (increase)
Ogura	12.74	13.36 "
Mitsubishi	8.36	8.90 "
Other Japanese	3.44	3.71 "

Source: Letter, F. W. Starling (Petroleum Department) to C. W. Orde (Foreign Office), September 8, 1934, FO371(1934)F5436/1659/23; see also letter, Kersey F. Coe (Stanvac, New York) to Hornbeck, August 28, 1934, RG 59, 894.6363/68; and letter, Grew to SecState, September 6, 1934, RG 59, 894.6363/78.

make matters worse, the quotas announced for the second half of 1934 clearly favored domestic companies and cast doubt on the future prospects of the two foreign companies in Japan. While permitted approximately the same actual volume as before, Stanvac and Rising Sun were not allocated any of the forecast increase in demand, and the percentage of the market allocated to them actually declined.[14] Protests produced an unequivocal reply from the Japanese Foreign Office on October 31, affirming that "it is Japan's desire in general to refine imported raw material rather than to import the manufactured article, and accordingly, in view of the fact that there are idle refining plants, it is intended in the first place to allot to them such proportion of increase in demand as is appropriate."[15]

Unlike the situation in Manchuria, no treaties had been violated, and there was no spontaneous reaction from the Anglo-American diplomatic corps. In fact, Washington and London did virtually nothing until prodded into action by the two oil companies. Stockpiling and restrictive quotas, however, struck at the heart of Stanvac and Rising Sun's sizeable Japanese investment, and they launched a long and acrimonious counterattack.[16] The record reveals a fairly straightforward and

[15]Enclosure in letter, Clive to Simon, November 9, 1934, FO371 (1934)F7160/1659/23.

[16]The Associated Oil Company (Tidewater Associated after 1936) owned half interest in Mitsubishi Oil, the third largest refiner in Japan, but its business was actually helped rather than hurt by the Petroleum Industry Law, and it never became involved in the protests made by Stanvac, Rising Sun, and the Anglo-American diplomatic corps. In 1923 Associated had entered into an exclusive sales agreement with Mitsubishi, and in 1929 the two concerns formed Mitsubishi Oil for the purpose of refining California crude in Japan, completing the first facility in 1931. When the new law required construction of extra storage facilities, Associated quietly advanced half the necessary capital—just over $200,000—and continued shipping an increasing volume of crude to Japan. By 1940 Tidewater Associated's total investment in Mitsubishi Oil amounted to $1,409,819, on which it had earned $521,364 between 1932 and 1940, an annual return of about five

successful course of action by the oil companies, who enlisted diplomatic support in just the proper degree to achieve results, but careful analysis reveals a wide divergence of motivation between oil men and diplomats. In effect, the oil companies used diplomatic concern over treaty violations in Manchuria to enlist governmental support for their business interests in Japan. As in Canton, institutional interests intersected for a brief period, and the groups gained additional experience acting as a team.

DESPITE forebodings during the spring and summer of 1934, the actual confrontation with the Japanese government built up slowly. The Ministry of Commerce and Industry first requested that Stanvac and Rising Sun file complete plans of operation by July 31, but intervention by Sir Robert Clive, the new British Ambassador to Tokyo, gave the companies more time to comply.[17] Both companies submitted tentative plans for the second half of 1934 by mid-summer, but were given until September 30 to prepare plans for 1935. In phraseology that turned out to be extremely important for subsequent negotiations, the companies were directed to include in these plans provisions for building up a three-month reserve stockpile *in addition to working stocks* by April 1, 1935, with an increase to a six-

---

percent in addition to *its* profits on the sale of crude. This position was quite different from that of Stanvac, which based its business on the direct sales of externally refined products, and it is hardly surprising that the two companies reacted differently. A full description of Tidewater Associated's Japanese business is contained in a letter from William F. Humphrey, its president, to Secretary of the Treasury Morgenthau, September 11, 1940, Morgenthau Diary, Vol. 306, pp. 185–91, Henry J. Morgenthau, Jr., MSS, Franklin D. Roosevelt Library, Hyde Park, New York, hereafter cited as "Morgenthau Diary." For Mitsubishi's position in the Japanese refining business, see attachments to letter, Morgenthau to Ickes, August 30, 1940, Vol. 292, pp. 231 ff, Morgenthau Diary.

[17]Telegram, Clive to Foreign Office, July 26, 1934, FO371(1934) F4575/1659/23; Foreign Office memorandum, August 8, 1934, FO371 (1934)F4994/1659/23.

month reserve by October 1, 1935.[18] While worrying over how to respond, Stanvac and Rising Sun submitted a long list of questions to the Ministry of Commerce and Industry, in effect asking for a full guarantee of their future status before deciding whether to make the necessary investment.[19] It was at this juncture that the issue exploded. On August 20 the Japanese government returned the original plans submitted by Stanvac and Rising Sun for the second half of 1934 with a directive that they be resubmitted by August 31 with *a reduced percentage of the market*.[20] The issue had been joined, and the reaction was instantaneous.

By sheer coincidence this was the precise moment when the embargo idea growing out of the Manchurian dispute reached London and Washington. It will be recalled that the idea originated when the Tokyo managers of Stanvac and Rising Sun suggested the *hint* of an embargo on crude as a psychological bargaining tool in *Manchuria* where British and American diplomats were deeply concerned over violations of the Open Door principle. Ambassadors Clive and Grew had passed the suggestion on to the Foreign Office and State Department with strong endorsements on August 17 and 20 respectively, and parallel communications went to the home offices of Stanvac and Asiatic in New York and London.[21] Also by coincidence, Deterding of Royal Dutch-Shell was in the United States, and he joined Teagle, president of Standard

[18]Memorandum of August 21, 1934, submitted to the Department of State by W. C. Teagle, RG 59, 894.6363/84; letter, Grew to SecState, August 24, 1934, RG 59, 894.6363/71.

[19]Letter, Grew to SecState, August 24, 1934, RG 59, 894.6363/71; telegram, Rising Sun (Yokohama) to Asiatic Petroleum (London), August 17, 1934, copy in FO371(1934)F5104/1659/23.

[20]Letter, Kersey F. Coe (Stanvac, New York) to Hornbeck, August 28, 1934, RG 59, 894.6363/68; letter, Grew to SecState, September 6, 1934, RG 59, 894.6363/78.

[21]Telegram, Clive to Foreign Office, August 17, 1934, FO371(1934) F5068/142/10; telegram, Grew to SecState, August 20, 1934. *FRUS (1934)*, III, 721–23; telegram, Rising Sun (Tokyo) to Asiatic Petroleum (London), August 17, 1934, FO371(1934)F5104/1659/23.

Oil (New Jersey) on a visit to Washington on August 22 to enlist governmental support. Experienced in working together on delicate matters, Teagle and Deterding deftly transferred the embargo idea from the Manchurian to the Japanese problem and came up with an ingenious scheme.

In meetings with Undersecretary of State William Phillips and Far Eastern Division Chief Hornbeck on August 22, and with Secretary of the Interior Harold Ickes and Hornbeck the following day, Teagle and Deterding proposed "frightening" the Japanese into moderation by hint of an embargo on the sale of crude oil to *Japan*.[22] They did not initially propose an actual embargo. Instead they suggested that the Department of the Interior, at the request of the State Department, conduct a survey of American petroleum exporters to determine the *feasibility* of a producers' agreement to restrict shipments to Japan. Such a move would obviously become known throughout the oil industry, and it would be timed to coincide with stiff diplomatic protests by both the British and American governments over the treatment being accorded Stanvac and Rising Sun in Japan. Although such a survey was never made, the embargo rumors generated by Teagle and Deterding's initiative may have produced exactly the results they intended.

The proposal requires a bit of explanation. As has been noted, Japan could meet less than ten percent of her petroleum requirements domestically, well over half of these requirements were procured from the United States, and most of the remainder came from the Netherlands East Indies. The Standard Oil Company (New Jersey) and Royal Dutch-Shell controlled not only the Indies output, but also a substantial

[22]Memorandum of conversation, Phillips and Hornbeck with Teagle and Deterding, August 22, 1934, RG 59, 894.6363/84; memorandum of conversation, Ickes, Hornbeck, and Dooman with Teagle and Deterding, August 23, 1934, RG 59, 894.6363/88; telegram, Phillips to Grew, August 31, 1934, *FRUS(1934)*, II, 728–29; Harold L. Ickes, *The Secret Diary of Harold L. Ickes* (3 vols.; New York: Simon and Schuster, 1953–54), I, 192. Ickes' diary entry suggests that he may have missed some of the subtlety of the proposal which appears in the longer State Department record of the meeting.

part of Middle Eastern production and some of that in the United States. Approximately half of Japan's imports, however, came from American companies that had absolutely nothing to gain from an embargo—notably Standard Oil of California, the Associated Oil Company, Union Oil, The Texas Company, and Richfield.[23] Although a portion of Japan's oil supply could be cut off by informal agreement between Stanvac's parent companies and Royal Dutch-Shell, overt action by some agency of the American government would be essential for an effective embargo.[24] What Teagel and Deterding initially proposed was a hint of such action to bring psychological pressure on the Japanese government. But just as was to be the case in a similar situation in 1940 and early 1941, the State Department proved extremely reluctant to sanction such a move. What the real reason for this reluctance was in 1934 is not at all clear from the available records, which are dominated by Hornbeck's attempts to evade either a firm decision or responsibility for indecision. The reluctance to take an overt step is consistent, however, with Dorothy Borg's conclusion that the State Department was profoundly shaken by the lack of worldwide support in the Manchurian crisis and studiously avoided overt antagonism of Japan during the rest of the 1930s.[25] A second possibility, which can only be surmised from

[23]A detailed, company by company, summary of the source of Japanese petroleum imports in the 1930s is included in a memorandum of August 6, 1940, provided to Secretary of the Treasury Morgenthau by Stanvac board chairman George Walden; Vol. 292, p. 268, Morgenthau Diary. For a summary of the 1939 data in that memorandum see Appendix B, Table B-6. General Petroleum also shipped to Japan, but has not been included in the list of independents because it was a wholly-owned subsidiary of Socony-Vacuum, the other parent of Stanvac (*Moody's Industrial Manual* 1970, p. 2655).

[24]As the embargo debate gained momentum, this key factor was discussed in considerable detail; see memorandum of conversation, Teagle with Hornbeck, Dooman, and Hamilton, September 13, 1934, RG 59, 894.6363/91, and letter, Edwin L. Neville (Chargé in Tokyo) to SecState, October 5, 1934, *FRUS(1934)*, III, 737–40.

[25]Borg, *The United States and the Far Eastern Crisis*, pp. 521–22. The reluctance to antagonize Japan over oil was also specifically noted

the surviving records, is that the State Department was aware that action to aid Stanvac would raise a furor among those oil companies benefitting from increased Japanese trade. A third possibility, supported somewhat by Hornbeck's ponderous evasions, is simply indecision in the face of a dilemma. In any event, the reluctance to act was consistent with Hull's basic policy of liberalizing international commerce and keeping *all* types of restraints on trade to a minimum. Whatever the reason, Teagle and Deterding met with cordial inaction by the American government, and the idea did not receive serious attention until it reached the British Foreign Office.

Before London picked up the embargo idea, however, events took a slightly more hopeful turn in Japan. On August 31, the Chief of the Mining Bureau of the Japanese Ministry of Commerce and Industry replied in writing to the extensive questions submitted by Stanvac and Rising Sun. Although the reply was by no means satisfactory it did include an invitation to discuss the issue further, and stated that "Unless something extraordinary takes place, the quantity [of petroleum] to be sold will not become less than the quantity sold at the time of the enforcement of the Petroleum Industry Law."[26] Stanvac and Rising Sun now had a minimum guarantee, and the door had been opened for what would prove to be a long period of negotiation. Informal contacts by company and diplomatic personnel at the Ministry of Commerce and Industry and the Foreign Office in September confirmed that key officials were in a slightly more conciliatory mood, partially because rumors of a possible embargo had reached the Japanese Foreign Office.[27] The two companies received additional support from

---

by Mira Wilkins in "The Role of U. S. Business" in Borg and Okamoto, *Pearl Harbor as History*, pp. 341–76.

[26]Letter, Grew to SecState, September 6, 1934, RG 59, 894.6363/78; the reply was also reported in a telegram, Clive to Foreign Office, September 21, 1934, FO371(1934)F5666/1659/23.

[27]Telegram, Clive to Foreign Office, September 21, 1934, FO371 (1934)F5666/1659/23; letter, Grew to SecState, September 21, 1934, RG 59, 894.6363/81.

formal overtures to the Japanese Foreign Office by Dutch Minister General Jean C. Pabst, on September 18, American Chargé Edwin Neville on September 25, and British Ambassador Clive on October 5.[28]

The Japanese demonstrated increased concern when Stanvac and Rising Sun refused to submit complete 1935 business plans by the September 30 deadline on the grounds that their future in Japan remained uncertain. The Commerce Ministry three times demanded immediate compliance,[29] and then on November 20 Vice Minister of Commerce Yoshino Shinji and Foreign Office Commercial Bureau Chief Kurusu Saburo met with Goold of Stanvac and Ely of Rising Sun to attempt to work out a compromise.[30] Yoshino and Kurusu offered the possibility of a ten-year guarantee of the present level of business and hinted that clear statements from the companies that the sheer cost of stockpiling could force withdrawal from Japan might help them persuade more intransigent factions in their government to adopt a less rigid position. They also indicated that they would accept provisional plans for 1935 "conditional

[28]Telegram, Grew to SecState, September 19, 1934, *FRUS(1934)*, III, 731–32; telegram, Clive to Foreign Office, September 21, 1934, FO371(1934)F5666/1659/23; telegram, Hull to Grew, September 21, 1934, *FRUS(1934)*, III, 733; telegram, Neville to SecState, September 25, 1934, *FRUS(1934)*, III, 737; telegram, Clive to Foreign Office, October 8, 1934, FO371(1934)F5999/1659/23; and telegram, Neville to SecState, October 10, 1934, *FRUS(1934)*, III, 740–41. The Japanese response on October 31 was moderate in tone (telegram, Grew to SecState, November 1, 1934, *FRUS(1934)*, III, 753–54).

[29]Letter, Grew to SecState, October 19, 1934, RG 59, 894.6363/ 107; telegram, Grew to SecState, October 22, 1934, *FRUS(1934)*, III, 743–44; telegram, Grew to SecState, November 6, 1934, *FRUS(1934)*, III, 754–55; telegram, Clive to Foreign Office, November 7, 1934, FO371(1934)F6661/1659/23; telegram, Clive to Foreign Office, November 14, 1934, FO371(1934)F6817/1659/23; letter, Grew to SecState, November 30, 1934, RG 59, 894.6363/153.

[30]Telegram, Clive to Foreign Office, November 21, 1934, FO371 (1934)F6932/1659/23; telegram, Grew to SecState, November 24, 1934, *FRUS(1934)*, III, 764–65; letter, Grew to SecState, November 30, 1934, RG 59, 894.6363/153; letter, Kersey F. Coe (Stanvac, New York) to Hornbeck, December 14, 1934, RG 59, 894.6363/152.

upon the foreign oil companies' decision as to whether or not it would be economically feasible for them to continue in business in Japan."[31] After consultation with their home offices, the American Embassy, the British Embassy, and the Netherlands Legation, Stanvac and Rising Sun agreed to submit provisional plans, and the tension momentarily subsided.[32] Grew reported on November 24 that Kurusu appeared ready to transfer the issue from diplomatic to private channels in order to work out a practical solution.[33]

The oil companies, while still desirous of diplomatic support for bargaining leverage, were of like mind. By early November Walden and Parker had concluded that the matter was sufficiently important to warrant personal intervention, and they decided to go to Tokyo along with F. Godber, a London director of Royal Dutch-Shell who was already planning a Far Eastern tour, for direct discussion with the Japanese government. The three would meet in Shanghai in mid-December for a major conference with their representatives in Japan and China, and wait there until diplomatic maneuvers had adequately paved the way.[34] As matters turned out, the Shanghai

[31]Letter, Grew to SecState, November 24, 1934, *FRUS(1934)*, III, 765.

[32]Telegram, Clive to Foreign Office, November 23, 1934, FO371 (1934)F6958/1659/23; telegram, Grew to SecState, November 24, 1934, *FRUS(1934)*, III, 766; telegram, Hull to Grew, November 28, 1934, *FRUS(1934)*, III, 770–71; and letter, Grew to SecState, November 30, 1934, RG 59, 894.6363/153.

[33]Telegram, Grew to SecState, November 24, 1934, *FRUS(1934)*, III, 764–65.

[34]Memorandum of conversation between Hornbeck, Teagle, and Parker, November 2, 1934, RG 59, 894.6363/112; memorandum, Phillips to Hull, November 9, 1934, Box 37, Folder 80, Cordell Hull MSS, Library of Congress, Washington, D.C., hereafter cited as "Hull MSS"; diary entry for December 1, 1934, Grew MSS; letter, Starling (Petroleum Department) to Randall (Foreign Office), December 3, 1934, FO371(1934)F7207/1659/23; telegram, Bingham (American Ambassador in London) to SecState, December 7, 1934, *FRUS(1934)*, III, 776–77; letter, Coe (Stanvac, New York) to Hornbeck, December 24, 1934, RG 59, 894.6363/157.

meeting took place as planned, but the assault on Tokyo was slightly delayed while the embargo proposal burned itself out in the British Foreign Office.

ONCE LAUNCHED by Teagle and Deterding, the embargo idea appears to have lived a life of its own, despite movement toward a negotiated *modus vivendi* in Japan. The reason is not hard to find. With minimal prodding from the oil companies, Japan and Manchuria became inextricably intertwined in diplomatic thinking at a time when anger was mounting over violations of the Open Door in the latter territory. While the State Department balked at taking overt steps toward an embargo, the Foreign Office had no such compunctions, possibly because any *overt* action would have to come from the American government. At any rate, the Foreign Office spent considerable effort attempting to generate support for a full-scale embargo on crude oil as psychological protection for British interests throughout East Asia. The attempt failed, but the rumors thereby generated appear to have had some moderating effect on the Japanese position. This threat could only remind the Japanese once again of their vulnerability in an increasingly hostile world.[35]

When Teagle and Deterding had originally seen Undersecretary of State Phillips and Secretary of the Interior Ickes, Hornbeck had suggested that since the British company had

[35]A trace of worry runs throughout Japanese official comments on petroleum during this period, but documents searched in American archives did not substantiate the author's original hypothesis that the 1934 embargo episode itself significantly influenced the Japanese drive for autonomy. The incident is not mentioned, for example, in the translated portions of the *Saionji Memoirs* or the *Kido Diary* (Harada Kumaō, *Saionji-Harada Memoirs* [1931–40], translation on reels SP 49–51; and *Kido [Kōichi] Diary* [1931–45], translation on reel WT 5; both in microfilms of documents from the Japanese Ministry of Foreign Affairs, 1868–1945, Library of Congress, Washington, D.C.). The episode and the reaction to it appear to have simply fit into the broader trends of the decade and certainly did nothing to disabuse the Japanese of their concern.

the larger stake in Japan any initiative toward an embargo should come from London, and Deterding had agreed to take the matter up immediately on his return.[36] The Foreign Office found, however, that the Admiralty had serious reservations, and the idea was allowed to gestate for two months while diplomatic discussions continued in Japan.[37] The gestation appears to have been more rapid in Tokyo than in London or Washington. On October 5, 1934, Edwin Neville, the American Chargé in Japan forwarded a long dispatch to Washington, arguing that "the oil problem in Japan and Manchuria . . . indicates gross disregard of the rights of and legitimate interests of other nationals, [and] is more than a mere matter of the protection of commercial interests . . . . Petroleum is Japan's weakest point. . . . By controlling the supply of crude oil at the sources . . . the interested countries may be able to induce a more reasonable attitude on the part of the Government and the people here toward other peoples and toward international relations in general."[38] Grew backed this up with a strong telegram on October 29 arguing that "the time has come to link together . . . the oil control system in Japan and the proposed oil monopoly in Manchuria."[39] Neville and Grew failed to move the Department of State,[40] but a telegram of similar intent

[36]Memorandum of conversation, Teagle and Deterding with Phillips and Hornbeck, August 22, 1934, RG 59, 894.6363/84; and memorandum of conversation, Teagle and Deterding with Ickes and Hornbeck, August 23, 1934, RG 59, 894.6363/88.

[37]Foreign Office minutes of October 26, 1934, FO371(1934)F6312/1659/23. The reference is to a meeting with Admiralty personnel on September 18.

[38]Letter, Neville to SecState, October 5, 1934, *FRUS(1934)*, III, 737–40.

[39]Telegram, Grew to SecState, October 29, 1934, *FRUS(1934)*, III, 750–51.

[40]Memorandum of Conversation, Hornbeck and Teagle, October 24, 1934, *FRUS(1934)*, III, 746; telegram, Phillips to Grew, October 31, 1934, *FRUS(1934)*, III, 752–53; Phillips told Grew that "We do not intend to be drawn or pushed into a position of taking the initiative in action or threats of action which . . . would lay us open to particularization of Japanese animosity . . . for reasons which need not be tele-

87

from Sir Robert Clive on October 24 reopened interest at the Foreign Office. Clive noted that "Japanese authorities insist on compliance by foreign companies of [the] provisional law requiring them to furnish at once their import sales and stock-keeping programme of 1935. Manchukuo authorities appear to be proceeding with their oil monopoly project. Critical stage in oil question has certainly now been reached . . . [and] to abstain from using in moderation our advantage is to make an unprofitable sacrifice in the cause of friendship."[41] Sentiment began to develop in London that "failure on our part to resist the Japanese pretensions in this affair is likely to have most unfortunate repercussions not only in Manchuria and Japan but also in China."[42] The two problems were now inextricably linked and rapidly becoming a major policy issue.

By October 30, 1934, Charles W. Orde, Head of the Foreign Office Far Eastern Department, became convinced that "the stiffest possible resistance should be offered both to the Japanese Petroleum Industry Law and the Manchurian monopoly," including governmental approval of a privately organized oil embargo and enlistment of American support.[43] This position was backed by the Petroleum Department,[44] but strongly opposed by the Board of Trade, which feared economic retaliation, and the Admiralty, concerned that an embargo might provoke "extreme Japanese action such as seizure

---

graphed but which have been explained here." This is one of the clearest statements made of the American position, but the rationale remains unclear. Hornbeck also told Parker that "the American Government awaits an initiative by the British Government and we definitely do not intend to 'get out in front' in this connection" (memorandum of conversation, Hornbeck and Parker, November 10, 1934; FRUS(1934), III, 757–58).

[41]Telegram, Clive to Foreign Office, October 24, 1934, FO371(1934) F6312/1659/23.

[42]Foreign Office minutes of October 26, 1934, FO371(1934)F6312/ 1659/23.

[43]Foreign Office memorandum of October 30, 1934, FO371(1934) F6426/1659/23.

[44]Letter, Starling to Randall, November 12, 1934, FO371(1934) F6771/1659/23.

of oil wells in the East Indies."[45] By this time events in both Japan and Manchuria had reached a tense phase, and on November 9 Orde recommended that the impasse within the British Government be resolved by a full Cabinet decision.[46] Accordingly, the Foreign Secretary, Sir John Simon, raised the issue at a Cabinet meeting on November 21, and the Foreign Office obtained approval for its course of action, provided American support could be ensured.[47] This was the catch—the Foreign Office had not read through American equivocation sufficiently to see that no overt support would be forthcoming. With the issue now squarely joined, however, the time for equivocation had passed.

On November 27 Orde arranged a conference with Hugh Millard, Second Secretary of the American Embassy, and Eugene H. Dooman of the State Department, who was in London with the American delegation to the preliminary conversations on the naval treaty then in progress.[48] Orde outlined British views with exceptional clarity. The British would approve a private embargo agreed upon by the oil companies, provided it was "watertight," did not require formal action by the British government, and would be applied to both Man-

[45]Foreign Office memorandum of October 30, 1934, FO371(1934) F6426/1659/23.

[46]Foreign Office memorandum of November 9, 1934, FO371(1934) F6727/1659/23.

[47]Cabinet paper on "Oil Position in Japan and Manchuria," November 20, 1934, FO899(1934), CAB 24/251, CP 262(34); Cabinet conclusion 41(34), November 21, 1934, FO899(1934), CAB 23/80, 41(34)4.

[48]The fact that rather tense discussions were in progress between the United States, Britain, and Japan on the possibility of a second renewal of the 1922 Washington Naval Treaty at the same time that the oil embargo was under consideration generated comment in a number of official documents, with repeated denial that there was any connection between the two (see, for example, telegram, Hull to Grew, November 28, 1934, *FRUS(1934)*, III, 771). On the naval talks, which ended with a formal Japanese announcement that the treaty would not be renewed when it expired in 1936, see Borg, *The United States and the Far Eastern Crisis*, pp. 106–12.

churia and Japan or not at all. What was the American position?[49] The query went immediately to Washington, and while
a reply was awaited Walden impatiently cabled from Shanghai
his desire to get on with commercial negotiations.[50] His request
was blocked by the Foreign Office on the theory that showing
"readiness to negotiate at the present juncture [might] be taken
for a sign of weakness."[51] In a particularly revealing memorandum, Orde had noted that premature settlement of the
companies' grievances in Japan would weaken British attempts
to bring Japan to heel in Manchuria—a simultaneous issue
was needed in both areas to justify an effective embargo.[52] The
British position collapsed completely, however, when Millard
and Dooman returned on December 7 with word from the
State Department. In tortuous but this time clear language,
Washington said: "We do not believe that . . . [without] definite
restrictive action on the part of the American Government effective restriction of petroleum exports from the United States
to Japan and Manchuria could be achieved. This government
does not for the present feel moved to proceed in the direction
of such action and it does not look as though the oil companies
adversely affected are in a position to take or cause the oil industry to take such cooperative action as might be effective."[53]
Since the United States did not "feel moved to proceed" toward
overt action, and since the companies were anxious to get on
with direct commercial negotiations, the Foreign Office bitterly
threw in the towel, and this ended the embargo debate of

[49]Memorandum of conversation, Orde and Randall with Millard
and Dooman, November 27, 1934, FO371(1934)F7079/1659/23;
telegram, Bingham to SecState, November 27, 1934, *FRUS(1934)*, III,
769–70.

[50]Telegram, Walden, in Shanghai [to Shell, London], December 5,
1934; copy in FO371(1934)F7283/1659/23.

[51]Foreign Office minutes of December 7, 1934; FO371(1934)F7283/
1659/23.

[52]Foreign Office memorandum of November 9, 1934, FO371(1934)
F6727/1659/23.

[53]Telegram, Hull to Bingham, December 5, 1934, *FRUS(1934)*,
III, 774–76.

1934.[54] After an extended exchange on how best to pave the way, American and British Embassy personnel met with Kurusu on December 27, and the Stanvac and Shell representatives were cleared to leave Shanghai and begin talks in Tokyo in early January.[55] The idea of curbing Japan by curtailing her oil supply was not seriously considered again until 1940.[56]

ONCE CLEARED to commence talks, the Stanvac and Shell dele-

[54]Memorandum of conversation, Orde and Randall with Millard and Dooman, December 7, 1934, with extensive minutes on the Foreign Office reaction; FO371(1934)F7300/1659/23.

[55]Telegram, Grew to SecState, December 27, 1934, *FRUS(1934)*, III, 798; telegram, Clive to Foreign Office, December 28, 1934, FO371 (1934)F7674/1659/23; telegram, Grew to SecState, December 29, 1934, *FRUS(1934)*, III, 798–99; telegram, Clive to Foreign Office, December 29, 1934, FO371(1934)F7702/1659/23.

[56]The published documents in the *Foreign Relations* series are a bit misleading on this point, since they could be read to imply that Shell reopened the whole embargo idea when the Manchurian monopoly went into effect in April 1935. Unpublished American documents and those in the British archives reveal that a full embargo was never contemplated. All that transpired was an unsuccessful attempt by Teagle, working through A. H. DeFriest of General Petroleum, to enlist the cooperation of California companies in withholding crude from the Manchurian refinery, for nuisance value only, since supplies could easily have been transshipped through Japan. Furthermore, Teagle was acting at the prompting of Hornbeck, who met him in New York on March 21, three weeks before a similar suggestion from Shell arrived through the British Ambassador in Washington. Hornbeck's motivation remains unclear, except possibly his perennial "let's you and him fight" attitude; memorandum of conversation, Hornbeck with Teagle, Dundas (Stanvac), and Coe (Stanvac), March 22, 1935, RG 59, 893.6363 Manchuria/155; memorandum of conversation, Hornbeck with Dundas, March 28, 1935, RG 59, 893.6363 Manchuria/158; Foreign Office memorandum of April 4, 1935, FO371 (1935)F2227/94/23; telegram, Hull to Grew, April 12, 1935, *FRUS (1935)*, III, 893; letter, Hornbeck to Lindsay (British Ambassador in Washington), April 16, 1935, *FRUS(1935)*, III, 900–903; telegram, Lindsay to Foreign Office, April 17, 1935, FO371(1935)F2546/94/23; memorandum of conversation, Mackay (State Department) with Coe, April 18, 1935, FRUS(1935), III, 903–904.

gations went on to Tokyo, completed coordination of their position with the British and American Embassies, were formally introduced to Japanese officials, and began serious negotiations on January 9, 1935.[57] Omens were good. Quotas for 1935, released in late December, gave both companies a share of the projected increase in consumption instead of the restrictive quotas set for the last half of 1934, and this change was taken as a conciliatory gesture.[58] To avoid excessive publicity and formality, meetings were held at the exclusive Tokyo Club, with no diplomatic personnel present. In addition to Parker, Walden, and Godber, the oil company representatives included John C. Goold, Stanvac's general manager in Japan, and H. W. Malcolm, managing director of Rising Sun. Kurusu and Yoshino, Chief of the Foreign Office Bureau of Commercial Affairs, and Vice Minister of Commerce and Industry respectively, represented the Japanese government, with official minutes taken by Kurusu's personal secretary.[59] As matters turned out, negotiations extended through thirteen meetings in January and February, a month-long adjournment in March, and four tense sessions in April. But persistence prevailed, and Parker, Walden, and Godber left Japan with what appeared to be a highly satisfactory agreement.

At the outset it became apparent that Kurusu and Yoshino

[57]Telegram, Grew to SecState, January 10, 1935, *FRUS(1935)*, III, 878; letter, Grew to SecState, January 10, 1935, RG 59, 894.6363/175; letter, Starling (Petroleum Department) to Randall (Foreign Office), January 14, 1935, FO371(1935)F321/94/23.

[58]Letter, Starling to Randall, January 1, 1935, FO371(1935)F95/94/23; memorandum, Department of Overseas Trade to Foreign Office, January 31, 1935, FO371(1935)F690/94/23; letter, Grew to SecState, February 8, 1935, RG 59, 894.6363/178.

[59]Letter, Grew to SecState, January 10, 1935, RG 59, 894.6363/173; Grew, *Turbulent Era*, p. 976; entry for January 10, 1935, Vol. 75, Grew Diary, Grew MSS. A complete set of English translations of the minutes, approved by Malcolm, survived with the records of the British Embassy in Toyko; file 269(1935), "Oil: Japan," British Embassy in Consular Archives: Japan, class FO262, Public Record Office, London; hereafter cited as "file 269(1935)FO262." Additional Japanese and company personnel occasionally joined in the meetings.

understood the practical problems of Stanvac and Shell and honestly sought a *modus vivendi* to avoid the complications of the companies deciding to withdraw from Japan. They appeared to be under considerable pressure, however, from more intransigent factions in the government, and when pressed by Parker and Godber on details, they repeatedly resorted to elusive generalizations.[60] Although pricing came up for considerable discussion, the real issues proved to be preferential quotas for domestic refiners over foreign importers and the onerous stockpiling regulations. When the discussions revealed that Kurusu and Yoshino were under exceptional pressure not to yield to foreign demands while the Diet was in session, Parker, the specialist in Asian psychology, suggested the subtle device of an adjournment until the end of the Diet session so that Kurusu and Yoshino could negotiate within the government without the appearance of foreign coercion.[61] To make this more plausible, Parker, Walden, and Godber agreed to leave Japan for a month and return the first of April. Accordingly, the discussions to date were carefully summarized as bargaining tools for Kurusu and Yoshino, and the three principal representatives of the companies left Tokyo on March 2 to conduct business in the Philippines.[62]

When the oil delegation returned to Japan, they found

[60]See minutes of meetings, January 9–February 22, 1935, file 269 (1935)FO262.

[61]Minutes of meeting of February 7, 1935, file 269(1935)FO262. The minutes clearly show that the idea of an adjournment originated with Parker, despite Grew's later report that it came from the Japanese; telegram, Grew to SecState, February 8, 1935, *FRUS(1935)*, III, 880; see also diary entry for February 19, 1935, Grew MSS.

[62]Memorandum of matters discussed between the Ministry of Commerce, the Standard-Vacuum Oil Company, and the Rising Sun Company, February 20, 1935, file 269(1935)FO262; and minutes of meetings, February 14, 19, 26, and 27, 1935, file 269(1935)FO262; a copy of the memorandum is also in a letter, Clive to Foreign Office, April 1, 1935, FO371(1935)F2124/94/23; Grew's report on the status of the memorandum was more pessimistic than the minutes would seem to warrant; telegram, Grew to SecState, March 2, 1935, *FRUS (1935)*, III, 883–84.

Yoshino in a stubborn frame of mind again—perhaps as a face-saving gambit. If the companies would not comply with the stockholding regulations, his Ministry would "naturally" have to give quota preference to those Japanese firms that did comply.[63] Stanvac and Shell, however, had prepared a counter-move which they advanced with infinite care in their April meetings. They had already decided to make no further investments in Japan, and if forced to comply with the stockpiling requirement, they were prepared to cut back sales so that present tankage would be the equivalent of six months stock. Since at that point Stanvac and Rising Sun's distribution network handled about forty-six percent of the gasoline sold in Japan, such a sudden reduction would cause considerable consumer unrest.[64] Kurusu and Yoshino caught the hint, but as a *coup de grace,* the oil companies enlisted the support of former American Ambassador to Japan, W. Cameron Forbes, who was in Tokyo as head of a National Foreign Trade Council delegation assessing market prospects in East Asia. Forbes spoke with a number of officials, including Yoshino, pointing out that excessive stubbornness on the petroleum law could adversely affect prospects for future Japanese-American trade.[65]

At this point, Yoshino and Kurusu yielded, agreeing on

[63]Minutes of meeting of April 5, 1935, file 269(1935)FO262.

[64]The full countermove is explained in a letter, Grew to SecState, April 5, 1935, RG 59, 894.6363/196; and the subtle use of this threat shows primarily in the minutes of the meeting of April 11, 1935, file 269(1935)FO262.

[65]Letter, Grew to SecState, April 19, 1935, RG 59, 894.6363/199; Grew, *Turbulent Era,* p. 977; diary entry for April 15, 1935, Grew MSS; minutes of meeting of April 10, 1935, file 269(1935)FO262; and telegram, Clive to Foreign Office, April 15, 1935, FO371(1935)F2458/94/23. The incident is also mentioned in the *Report of the American Economic Mission to the Far East: American Trade Prospects in the Orient* (New York: National Foreign Trade Council, 1935), p. 12; the Standard-Vacuum Oil Company had been one of fourteen American companies financing the Mission, and John Goold, general manager of its Japanese office, was among the official hosts for the group in Tokyo. Forbes came away with the impression that his intervention had been decisive, but that appears questionable in view of all the

April 13 to a "five-point memorandum" which for the next three years the companies construed to have all the effect of a treaty. Stanvac and Rising Sun were guaranteed their 1935 gallonage as a minimum, plus at least one-third of future increases, and a price structure that would not force them to sell at a loss. The three-month stockholding requirement, which had become effective April 1, would be interpreted to include *working stocks already on hand,* and the six-month requirement would be dropped before it became effective on October 1.[66] Since the companies already carried roughly a two-month reserve in working stocks, this arrangement effectively permitted them to continue operation in Japan with minimum additional expense and inconvenience. In Grew's opinion the Japanese negotiators had wanted such a settlement all along, but needed a show of strength by the oil companies to convince their more intransigent colleagues that they had no alternative but to yield. Kurusu and Yoshino were later to maintain that the memorandum represented a statement of what they would *like* to accomplish, but Stanvac and Shell held for three years that it had constituted a firm commitment. For the time being at least, Parker, Walden, and Godber had accomplished about all they could, and they left Tokyo well pleased.

AGREEMENTS made under pressure have a way of coming unstuck once the pressure diminishes, and the five-point memorandum proved no exception. Until the beginning of the undeclared war in China in 1937, the Japanese government made repeated attempts to force Stanvac and Rising Sun to

---

other pressures at work (Forbes Journal, Volume 10, entries for April 9, 1935, p. 158; April 12, 1935, pp. 163–64; June 2, 1935, p. 247; June 8, 1935, p. 259; and June 21, 1935, p. 279; W. Cameron Forbes MSS, Houghton Library, Harvard University, Cambridge, Massachusetts).

[66]Telegram, Grew to SecState, April 13, 1935, *FRUS(1935)*, III, 896; "Memorandum by the Foreign Oil Interests in Japan," *FRUS (1935)*, III, 896–97; telegram, Clive to Foreign Office, April 13, 1935, FO371(1935)F2449/94/23; letter, Clive to Simon, April 16, 1935, FO371(1935)F3345/94/23; letter, Grew to SecState, April 19, 1935, RG 59, 894.6363/199.

stockpile at their own expense. The companies countered with threats to withdraw, offers to sell a hydrogenation process to Japan, prolonged negotiations with Mitsui to carry the required reserve stocks for them, and a resurrection of diplomatic pressure based on the Manchurian claim. As a result, the impasse continued well into 1937, when the Japanese economy began shifting to a full wartime orientation, and emphasis changed to procuring all the oil that could be obtained with minimum alienation of suppliers.[67]

As early as July 1935, both Grew and Clive reported that Kurusu and Yoshino were having difficulty convincing other departments—especially the Army and Navy—that the six-month stockholding requirements should be dropped as promised before it went into effect on October 1.[68] Before this could be resolved, however, domestic problems intervened. Japanese refiners were preparing to comply with the law, but they complained about the extra expenditure. They had not been given permission to increase prices, since such a move would create strong consumer resentment. Direct governmental compensation for the extra expenditure was a possible solution, but such a plan would require time to develop. Largely because of these domestic complications, implementation of the six-month stockholding requirement was postponed by Imperial Ordinance of September 18 until June 30, 1936 for those companies requesting it.[69] Through no effort of their

[67]Although documents in the Anglo-American archives make these extended negotiations appear rather tense, the possibility exists that from the Japanese viewpoint this was only a face-saving method of permitting noncompliance.

[68]Telegram, Grew to SecState, July 18, 1935, *FRUS(1935)*, III, 922–23; letter, Clive to Sir Samuel Hoare (Foreign Secretary), August 3, 1935, FO371(1935)F5485/94/23; memorandum of conversation, Grew with Kurusu, July 18, 1935, Vol. 1, pp. 195–96, Grew MSS.

[69]Letter, Clive to Hoare, August 6, 1936, FO371(1935)F5936/94/23; letter, Grew to SecState, September 6, 1935, RG 59, 894.6363/209; telegram, Neville to SecState, September 28, 1935, *FRUS(1935)*, III, 926–27; Economic and Trade Notes No. 31 of September 18, 1935, and No. 37 of October 4, 1935, by Paul P. Steintorf, Box 1528, "Tokyo" folder, RG151; and Grew, *Turbulent Era*, pp. 977–78.

own Stanvac and Shell received a nine-month reprieve from another direct confrontation.

Pressure did not abate, however, because domestic companies unexpectedly began to comply with the original deadline, and Kurusu and Yoshino found it necessary to shift their position.[70] On November 6 they told the local Stanvac and Rising Sun managers that the five-point memorandum had represented only what they had hoped to accomplish, and since it had been found impossible to permanently change the six-month stockholding requirement, the companies would be expected to comply by next June 30. Stanvac and Rising Sun had already been informed that a plan was being developed to provide a subsidy in the form of a six percent return on capital invested in extra tankage and stocks, and in the meantime those Japanese companies that had built up stocks would be compensated by giving them most of the quota for the projected 1936 increases in the market.[71] This position sent a shock wave through the oil companies and the Anglo-American diplomatic establishment, and generated a flurry of telegrams and meetings to decide on a counter strategy.[72] Again, Parker had a card to play—this time a controversial one. Since the Japanese were so concerned about oil reserves, why not offer to provide them with rights to the coal hydrogenation process

[70]Telegram, Clive to Foreign Office, October 9, 1935, FO371(1935) F6391/94/23; Japanese companies were apparently influenced by wording in the new regulations that required formal application for postponement, effectively admitting that they had not been prepared to comply.

[71]Telegram, Neville to SecState, November 15, 1935, *FRUS(1935)*, III, 930–31. The compensation and quota aspects of the position were presented by Yoshino and Kurusu on September 27 and reported in a letter, Neville to SecState, October 4, 1935, RG 59, 894.6363/213; and in a telegram, Malcolm (Yokohama) to Asiatic Petroleum (London), September 30, 1935, copy in FO371(1935)F6321/94/23.

[72]See, for example, minutes of a meeting between State Department and Stanvac personnel on November 6, 1935, RG 59, 894.6363/221; and minutes of a similar meeting of Foreign Office, Petroleum Department, and Shell personnel on November 15, 1935, FO371(1935) F7183/94/23.

jointly owned by Standard of New Jersey and Royal Dutch-
Shell—in exchange for a firm commitment to the five-point
memorandum?[73] Hornbeck was thoroughly shaken, and with
the personal concurrence of Secretary of State Hull, completely
disassociated the American government from any part in the
scheme. He did not, however, go so far as to tell Stanvac *not*
to make the offer.[74] The Foreign Office showed similar un-
easiness over the potential military implications, but offered no
objections.[75] Accordingly, overtures were made by company
representatives in Japan to Yoshino on December 14 and 23
and Kurusu on December 27, only to find them not interested.
Japanese research on converting coal to oil was believed to be
sufficiently far advanced and capital so fully committed that a

[73]Memorandum of conversation, Hornbeck, Dooman, and Mackay
with Parker, Walden, and Goold, November 6, 1935, RG 59, 894.6363/
221; and memorandum of conversation, A. Gascogne (Foreign
Office) with Sir Andrew Agnew (Shell), December 4, 1935, FO371
(1935)F7557/94/23. This was I. G. Farben's Bergius coal hydrogenation process, for which Jersey and Shell held worldwide rights
outside of Germany through International Hydrogenation, Ltd., incorporated in Liechtenstein (Larson, Knowlton, and Popple, pp. 153–
59). Despite the Japanese 1935 profession of confidence in their own
research, Mitsui negotiated directly with International Hydrogenation
in 1936 but ended by acquiring rights to the German Fischer-Tropsch
process from Ruhrchemie, A. G.. Mitsubishi tried throughout World
War II to obtain use of I. G. Farben's Bergius process and did not
succeed until January 1945. Japanese research had been conducted
primarily by the Navy, which had built a shale oil plant at Fushun,
Manchuria. None of these projects came close to meeting Japanese
expectations in World War II. See Boyce (American Consul at Yokohama) to SecState, May 9, 1936, RG 59, 894.6363/276; Grew to Sec-
State, December 15, 1936, RG 59, 894.6363/303; and Jerome B.
Cohen, *Japan's Economy in War and Reconstruction* (Minneapolis:
University of Minnesota Press, 1949), pp. 138–39.

[74]Memorandum of December 11, 1935, conversation between
Hornbeck and Parker, RG 59, 894.6363/230; telegram, Hull to Grew,
December 13, 1935, *FRUS(1935)*, III, 936.

[75]Foreign Office minutes of December 12, 1935, with comments by
Gascogne and Orde, FO371(1935)F7787/94/23.

new process was not needed.[76] The offer was left open, but for once Parker had misjudged his opposition.

The next suggestion, oddly enough, came from Kurusu, and provoked a brief rift in the Anglo-American front. At a meeting on December 27, Kurusu suggested that Stanvac and Rising Sun look into some type of business arrangement with a Japanese company such as Mitsui Bussan Kaisha to permit technical compliance with the law.[77] For a number of years Mitsui had handled bulk sales for Stanvac outside of the American company's own distribution network in Japan, and Parker showed interest. But Sir Andrew Agnew, who by 1936 had replaced Deterding in directing Royal Dutch-Shell's Asiatic affairs, disagreed and argued for a firm threat to withdraw unless the five-point memorandum was reaffirmed.[78] When Hornbeck learned of the split, he appealed directly to Teagle for a united position, and Teagle succeeded in welding the companies back together. After extended discussion an arrangement with Mitsui appeared the best alternative, and in March 1936 Stanvac and Rising Sun's new Japanese managers, C. E. Meyer and T. G. Ely, broached the subject first with Kurusu and Yoshino and then with Managing Director Mukai Tadaharu of Mitsui

[76]Telegram, Asiatic Petroleum (London) to Rising Sun (Yokohama), December 9, 1935, copy in FO371(1935)F7787/94/23; telegram, Stanvac (New York) to Stanvac (Yokohama), December 10, 1935, copy in RG 59, 894.6363/233; letter, Grew to SecState, December 14, 1935, RG 59, 894.6363/233; letter, Grew to SecState, December 14, 1935, RG 59, 894.6363/242; telegram, Stanvac (Yokohama) to Stanvac (New York), December 24, 1935, copy in RG 59, 894.6363/240; telegram, Grew to SecState, December 26, 1935, FRUS(1935), III, 938–39.

[77]Telegram, Rising Sun (Yokohama) to Asiatic Petroleum (London), December 27, 1935, copy in FO371(1935)F8137/94/23; letter, Grew to SecState, February 20, 1936, RG 59, 894.6363/255.

[78]Letter, Agnew to Orde, February 6, 1936, FO371(1936)F758/105/23; memorandum of conversation, Hornbeck with Parker and Goold, February 10, 1936, RG 59, 894.6363/250; and memorandum of conversation, Hornbeck and Mackay with Teagle and Parker, March 19, 1936, RG 59, 894.6363/265.

Bassan Kaisha. If an agreement could be reached, Mitsui would construct tankage and store fuel for the account of both Stanvac and Rising Sun, with compensation to come partly from the six percent governmental subsidy and partly from commissions on an additional allocation of Stanvac products for sale.[79] This would meet the intent of the law without the investment of additional Anglo-American capital in Japan. A pact had almost been consummated by June, permitting the two companies to avoid a crisis on the June 30 deadline by submitting a statement that "the companies are now negotiating a special arrangement to enable them to fulfill their obligations under the petroleum law and have reached an agreement in principle." Yoshino agreed to renew the companies' full quotas and not invoke the penalty section of the law.[80] But when the companies finally asked the Commerce Ministry to confirm in writing the fact that the proposed plan would be acceptable to the government, they were turned down, and negotiations with Mitsui ended in November.[81] In 1936 it was difficult to find any

[79]Memorandum of conversation, Hornbeck with Parker, February 17, 1936, RG 59, 894.6363/251; memorandum of conversation, Hornbeck with Teagle, February 25, 1936, RG 59, 894.6363/248; telegram, Asiatic Petroleum (London) to Rising Sun (Yokohama), March 19, 1936, copy in FO371(1936)F1597/105/23; memorandum of conversation, Hornbeck and Mackay with Teagle and Parker, March 19, 1936, RG 59, 894.6363/265; telegram, Grew to SecState, March 23, 1936, *FRUS(1936)*, IV, 792; telegram, Rising Sun (Yokohama) to Asiatic Petroleum (London), March 24, 1936, copy in FO371(1936) F1674/105/23; memorandum of conversation, Meyer (Stanvac) and Ely (Rising Sun) with Nanjo and Tajima (Mitsui; Mukai did not attend the first meeting, but handled all subsequent negotiations), March 26, 1936, copy in RG 59, 894.6363/271; telegram, Rising Sun (Yokohama) to Asiatic Petroleum (London), March 27, 1936, copy in FO371(1936)F1774/105/23.

[80]Letter, Grew to SecState, May 27, 1936, RG 59, 894.6363/277; telegram, Grew to SecState, June 23, 1936, *FRUS(1936)*, IV, 796–97; letter, Grew to SecState, July 10, 1936, RG 59, 894.6363/287; letter, Starling to Orde, October 2, 1936, FO371(1936)F6040/105/23.

[81]Letter, E. R. Dickover (Chargé in Tokyo) to SecState, October 14, 1936, RG 59, 894.6363/294; letter, Parker to Hornbeck, October 20, 1936, *FRUS(1936)*, IV, 798; telegram, Grew to SecState, Decem-

Japanese official who would formally approve a compromise with a foreign concern, and another confrontation appeared imminent.

With all other resources exhausted, the time had arrived for another barrage of diplomatic artillery. At the request of the two companies, Ambassadors Grew and Clive called on Foreign Minister Arita Hachirō on December 23 to register concern over the failure of the Commerce Ministry to provide written approval of an arrangement with Mitsui and to express the hope that Stanvac and Rising Sun would not be forced to retire from the country. This appeared to have a temporarily sobering effect, and the companies entered 1937 in technical violation of the law but with no penalties invoked by the Japanese.[82] It was at this point that Parker decided to resurrect the Manchurian claim, in order to "give the company additional leverage in connection with its negotiations relating to business in Japan."[83] Presumably he hoped to resurrect charges of treaty violations in Manchuria to embarrass the Japanese and make them more amenable to commercial concessions in Japan. At any rate, Stanvac and Asiatic began discussions with the Manchukuo government, and Grew added diplomatic support with a call on the Japanese Foreign Ministry on July 2, 1937.[84]

---

ber 14, 1936, *FRUS(1936)*, IV, 799–800; letter, Starling to Orde, December 19, 1936, FO371(1936)F7853/105/23; letter, Grew to SecState, April 2, 1937, *FRUS(1937)*, IV, 726–29.

[82]Memorandum of conversation, Hornbeck *et al.*, with Kersey F. Coe (Stanvac), December 17, 1936, RG 59, 894.6363/297; telegram, Clive to Foreign Office, December 17, 1936, FO371(1936)F7769/105/23; telegram, Grew to SecState, December 24, 1937, *FRUS(1936)*, IV, 805–806; telegram, Grew to SecState, January 11, 1937, *FRUS (1937)*, IV, 724–25; telegram, Grew to SecState, January 15, 1937, *FRUS(1937)*, IV, 725–26; and letter, Grew to SecState, April 2, 1937, *FRUS(1937)*, IV, 726–29.

[83]Memorandum of conversation, Hornbeck *et al.*, with Parker and Marshall (Stanvac), April 15, 1937; RG 59, 893.6363 Manchuria/294.

[84]Telegram, Grew to SecState, May 24, 1937, *FRUS(1937)*, IV, 731; telegram, Grew to SecState, July 2, 1937, *FRUS(1937)*, IV, 733. No

A new round of oil talks had started, but five days later fighting between Japanese and Chinese troops at the Marco Polo Bridge just outside Peiping began a process of military escalation that rapidly became full-scale undeclared war between Japan and China. As international tension mounted, Japan accelerated the process of placing her economy on a full wartime basis, and on August 19, 1937, the head of the Fuel Bureau in the Commerce Ministry urgently requested that the companies meet with him to resolve all outstanding differences.[85] This development effectively ended the debate over company-financed stockpiling; from that point on the Japanese concentrated on acquisition of reserves with minimum alienation of suppliers. Desultory talks continued for another year, but the Manchurian claim remained unsettled, the companies refused to stockpile at their own expense, and the Japanese took no further action to prevent continuation of their operations in Japan. The limited statistical data available do not reveal any drastic curtailment of Stanvac sales in Japan despite all the rhetorical thunder of the mid-1930s.[86] Three years of negotiations effectively ended in a stalemate, but by that time

---

evidence could be found in either the American or British records that the British government took similar diplomatic action at this point. It appears that a truce on the oil issue had begun to develop even before the Marco Polo Bridge incident.

[85]Telegram, Grew to SecState, August 20, 1937, RG 59, 894.6363/322; letter, Coleman (Petroleum Department) to Henderson (Foreign Office), August 23, 1941, FO371(1937)F5652/66/23; letter, Starling (Petroleum Department) to Ronald (Foreign Office), October 24, 1937, FO371(1937)F8646/66/23.

[86]Compare Table II-1 with Table B-1, Appendix B. From 1939 on the impact of Japanese regulations was obscured by other factors, including a shortage of tankers and foreign exchange and turmoil in the oil business created by the war in Europe. A position paper prepared in the British Foreign Office August 26, 1940, affirmed that Stanvac and Rising Sun were granted essentially constant import quotas from 1934 on, with Japanese companies dividing up the market increase (FO371(1940)W9865/9160/49).

the Sino-Japanese war had created new problems to be dealt with on the Chinese mainland.

IN RETROSPECT, it is clear that the teamwork already apparent in Canton in 1933 matured considerably in the face of Japanese attempts to achieve a higher degree of autonomy in oil in the period immediately before the Sino-Japanese war. For somewhat different reasons, Stanvac, Shell, the Department of State, and the Foreign Office worked together in countering Japanese moves with increasing smoothness and appreciation of one another's viewpoint. The oil companies consistently concentrated on keeping the Japanese market open with a minimum of unremunerative new investment, using tactics largely devised by Parker. Support could be rallied from the Anglo-American diplomatic establishment whenever treaty violations were invoked, but the State Department steadfastly balked at overt action that might provoke retaliation by Japan. In London, the Foreign Office took a more belligerent stance despite a cautious Admiralty, but ultimately had to defer to the United States since that was where control over Japan's oil supply actually lay. This general pattern of relationships persisted until mid-1941, and the teamwork evident in this earlier period proved immensely useful in evolving a highly sophisticated method for Anglo-American control of Japan's oil supply in the critical months of 1940 and 1941. The episodes in Manchuria and Japan provided additional impetus to the deterioration of Japanese-American relations, but ultimately the development of an effective team may have been the most significant by-product of the two incidents.

FOUR

CONFLICT IN CHINA

THE UNDECLARED Sino-Japanese war, which began in the summer of 1937, sharply increased the friction between Japan's drive for autonomy and America's Open Door policy, and Stanvac again found itself caught in the vortex of the conflict. With a distribution network spread throughout the area where most of the fighting took place, Walden and Parker went into an almost full-time alliance with Shell, the Department of State, and the Foreign Office to cope with repeated Japanese violations of treaty rights. The teamwork that began in the early 1930s in response to the Manchurian monopoly and the Petroleum Industry Law thus continued, with earlier differences of opinion muted as oil men and diplomats began to perceive the East Asian problem in much the same manner. Business interests and treaty rights became so intertwined as to be inseparable, but there was a shared reluctance to face a real showdown in China for fear it would precipitate a Japanese move against the far more critical resources of the Netherlands Indies.

The fighting that created Stanvac's problem in northeast China, although it was the culmination of a decade of mounting tension, was actually unplanned. Japanese forces had maintained constant pressure on North China since the Tangku Truce of May 1933, but there is no evidence that the skirmish with Chinese troops near the Marco Polo bridge on July 7, 1937, was part of a premeditated plan to launch a full-scale attack.[1] The Japanese units involved came from a 7,000 man

---

[1] This interpretation is based on Crowley, *Japan's Quest For Autonomy*, pp. 322–342; and Borg, *The United States and the Far Eastern Crisis*, pp. 276–317, 399–441.

detachment stationed near Peiping under the Boxer Protocol, and the clash occurred during routine field maneuvers. The incident had actually been resolved by agreement between local military commanders when pent-up tensions in Nanking and Tokyo triggered a process of escalation that transformed the incident into the long and bitter Sino-Japanese War. Specifically, Chiang Kai-shek ordered four divisions into combat readiness positions south of Peiping, the government of Prince Konoe Fumimaro began mobilizing additional forces in Japan for possible dispatch to the area, another local clash occurred on July 27, and a Chinese air attack on the Japanese navy installation at Shanghai on August 14 converted the incident into full-scale war. Chiang had finally resolved not to yield another inch of Chinese territory, the Konoe cabinet had become convinced that the Chinese army must be chastised to protect Japan's position on the continent, and repeated tactical victories drew Japanese forces further and further into the Chinese quagmire. Roosevelt signaled American concern with his famous "Quarantine Speech" in Chicago, but took no further unilateral action. China appealed to the League of Nations, which offered the opinion that Japan had violated the Nine-Power Treaty of 1922 and recommended that a conference of signatories of that treaty be convened in Brussels on November 3. Japan refused to attend, Germany declined a special invitation, Russia accepted a similar invitation but demonstrated caution on most concrete proposals, the United States and the smaller powers resisted serious discussion of economic sanctions, and Britain was reluctant to move out ahead of the United States. The conference settled on a mild reprimand to Japan, went into indefinite recess on November 24, and never reconvened. The Western powers did not have the means, inclination, or unity to intervene effectively.

Stanvac found itself caught in the center of military, political, and economic disruption which now spread throughout northeast China. Most fighting took place in the territory of the company's North China Division, which was based in Shanghai and responsible for marketing in all of China north of Foo-

chow.[2] Branch offices and storage facilities existed in most
major cities, including Peiping, Tientsin, Chefoo, Tsingtao, and
Tsinan in the North, and Nanking, Kiukiang, Hankow, and
Chungking on the Yangtze River. Suboffices were spread
throughout the region, but Chinese merchants handled most
of the actual retail sales.[3] The Division also operated a fleet of
thirteen small tankers on the Yangtze, the three largest of
which were the ill-fated *Mei Ping, Mei Hsia,* and *Mei An.*
Total investment in China, including facilities of the South
China Division based in Hong Kong, was approximately $8.8
million, slightly more than $7.7 million invested in Japan,
but annual sales were substantially less than those in Japan, as
shown in Table IV-1:

TABLE IV — 1

STANVAC 1937 SALES IN CHINA AND JAPAN
(In thousands of 42-gallon barrels)

	China	Japan
Kerosene	1,390	349
Diesel Oil	630	2,359
Gasoline	527	1,491
Fuel Oil	143	1,183
Lubricating Oil and Misc.	229	208
Total	2,919	5,590

SOURCE: Standard Oil Company (New Jersey), See Table B-1,
Appendix B.

An operation so thoroughly integrated into the economic life
of China could scarcely escape disruption as the war spread

[2] This description of Stanvac's China facilities is based on data assem-
bled in Appendix A.

[3] A number of British and American companies in China used this
type of distribution system, handling merchandise all the way to a cash
sale to the small local merchant; Edith E. Ware, *Business and Politics
in the Far East* (New Haven: Yale University Press, 1932), pp. 68–82.

south, and again Asiatic Petroleum found itself in a parallel position, with a similar though larger marketing organization in the same territory.

THE THREAT of Japanese expansion to this network was graphically driven home on December 12, 1937, when Japanese aircraft sank the *Mei Ping, Mei Hsia,* and *Mei An,* along with the American gunboat *Panay,* on the Yangtze River upstream from Nanking.[4] Japanese troops had been closing in rapidly on that city, Ambassador Johnson had moved to Hankow along with the Chinese Foreign Ministry, and the commander of the Yangtze Patrol had ordered the *Panay* to stay as long as possible to serve as a communication link with the small Embassy detachment still there and to take out any of the few remaining Americans who decided to depart at the last minute. On December 11, heavy shelling of the area finally forced the *Panay* to take the Embassy detachment aboard and leave, temporarily anchoring upriver out of artillery range. Accompanying the gunboat were the three small Stanvac tankers, carrying an estimated 800 Chinese employees and their families who were being evacuated from the city. The tankers were actually riverboats, resembling large oceangoing tugs in configuration, the *Mei Ping* at 1,118 gross tons, the *Mei Hsia* at

[4]Except where otherwise noted, this account of the *Panay* incident is based on the report of Second Secretary George Atcheson to SecState, December 21, 1937, *FRUS(Japan, 1931–41),* I, 532–41; findings of the Naval Court of Inquiry forwarded by the Commander in Chief, U.S. Asiatic Fleet (Yarnell) to the Secretary of the Navy (Swanson), December 23, 1937, *FRUS(Japan, 1931–41),* I, 542–47; and Hamilton D. Perry, *The Panay Incident: Prelude to Pearl Harbor* (New York: Macmillan, 1969), which adds considerable detail from the reminiscences of survivors. The episode is also recounted in Manny T. Koginos, *Panay Incident: Prelude to War* (Lafayette: Purdue University Press, 1967). There are only a few pieces of correspondence with Stanvac on this subject in State Department files, and most of these deal with subsequent damage claims. Presumably the company had nothing of significance to add to the instantaneous forceful protests by the American government.

1,048 gross tons, and the *Mei An* at 934 gross tons.[5] Each carried the distinctive Standard Oil "S" on its stack, each flew a large American flag, and each had an American flag painted horizontally on its superstructure or awning. Crews of all three were Chinese, except for the captains, who were European or American. The vessels had taken aboard several members of Stanvac's Nanking staff and as many Chinese employees as possible, including a number of blue-uniformed company security guards. Along with four other small Stanvac craft they had moved upstream to wait out fighting in the city, staying close to the *Panay* for extra protection.[6]

Shortly after 1:30 p.m. on Sunday, December 12, three twin-engine Japanese aircraft passed over the convoy at high altitude, releasing bombs which struck both the *Panay* and *Mei Ping*. Shortly thereafter, six single-engine aircraft dive-bombed and strafed the American vessels for twenty minutes, leaving the *Panay* sinking in midstream, the *Mei An* beached and burning downstream, and the *Mei Ping* and *Mei Hsia* ablaze and pulled alongside a small pontoon wharf on the south shore. Although the *Panay* had returned fire with 30-caliber machine guns, there was no apparent damage to the Japanese planes. Casualties were heavy, with two American sailors, an Italian newsman, and Captain Carlson of the *Mei An* all mortally wounded. Although available records do not include a casualty count among the Chinese refugees, the rapid destruction of the Stanvac vessels suggests that it must have been high. Also lost were two of the small Stanvac craft, with two others damaged. Survivors of the *Panay* crew, the four-

[5]Tōa kenkyūjo, *Shogaikoku no tai-Shi tōshi* [East Asia Research Institute, *Foreign Investments in China*] (3 vols.; Tokyo, n.p., 1942–43), uncatalogued copy in the Japanese Section, Orientalia Division, Library of Congress; relevant portions translated for the author by Kook-jin Rhee of Cincinnati, hereafter cited as "East Asia Research Institute," II, 590–91.

[6]Perry, pp. 60–61, 247. It is possible that there were more than four small craft, but that is the number given in the damage claim submitted to the Japanese government (telegram, Hull to Grew, April 7, 1938, *FRUS(Japan, 1931–41)*, I, 561–62).

man Embassy detachment, and nine other civilians made it ashore on the north bank, reached Ambassador Johnson in Hankow by telephone the following day, and were picked up by H.M.S. *Bee,* H.M.S. *Ladybird,* and U.S.S. *Oahu* on the 14th for return to Shanghai. Although premeditation cannot be completely ruled out, subsequent investigations strongly suggest that the incident was the result of communications confusion, lack of adequate preventive orders, and overzealousness on the part of the Japanese naval flyers involved.

Be that as it may, the impact was severe when news of the attack reached Washington.[7] Roosevelt immediately asked Hull to inform the Japanese Ambassador that he was "deeply shocked" and expected an "expression of regret and . . . full compensation."[8] Apologies were offered promptly and profusely by Foreign Minister Hirota in Tokyo and Ambassador Saito in Washington, and an offer of full compensation followed quickly. Washington and London conferred briefly on the possibility of even stronger action, and Roosevelt broached the idea of economic retaliation to Secretary of the Treasury Henry Morgenthau, Jr., who enthusiastically drafted an elaborate plan to freeze Japanese assets in the United States. By the time Morgenthau completed work on the proposal, however, the President's temper had cooled, and the Treasury plan was pigeonholed for another year.[9] Meanwhile, Japan had sent another profound apology, which the United States formally accepted on December 25.[10] The Department of State presented a claim for $2,214,007 for lives and property lost, of

[7]Except as otherwise noted, this account of reaction to the *Panay* incident is based on Borg, *The United States and the Far Eastern Crisis,* pp. 486–518.

[8]Memorandum by Hull, December 13, 1937, *FRUS(1931–41),* I, 522–23.

[9]John M. Blum, *From the Morgenthau Diaries* (3 vols.; Boston: Houghton Mifflin, 1959–67), Vol. I: *Years of Crisis, 1928–38,* pp. 485–92.

[10]Telegram, Grew to SecState, December 24, 1937, *FRUS(1931–41),* I, 549–51; telegram, Hull to Grew, December 25, 1937, *FRUS(1931–41),* I, 551–52.

which $1,287,942 represented Stanvac losses. Payment of this by the Japanese government on April 22 officially ended the episode.[11] Tempers had flared at this flagrant attack on American nationals and property, but apologies had been profuse, and sentiment in the United States clearly would not support a rashly belligerent reaction to events so far away.[12]

FROM STANVAC'S viewpoint, however, the sinking of three tankers was only the beginning of a long series of harassments to its China trade. Far more serious problems stemmed from progressive Japanese attempts to incorporate Occupied China into its own economic sphere, to the exclusion of all foreign interests. Although extreme militarists and established industrial interests had no abiding love for one another, they were both, after all, Japanese, and a wary cooperation developed in the territory seized by Japanese armies.[13] One of Stanvac's first serious encounters with this covert cooperation came when the Japanese military refused to reopen the Yangtze to commercial shipping long after the fighting had passed Nanking, on the pretext that the river was still mined, blocked with debris, and threatened by guerilla forces. It soon became evident,

[11]Telegram, Hull to Grew, April 7, 1938, *FRUS(1931–41)*, I, 561–62; telegram, Grew to SecState, April 22, 1938, *FRUS(1931–41)*, I, 563.

[12]A significant byproduct of the *Panay* incident was the impetus it gave to action on the Ludlow Amendment, which would have modified the Constitution to require a national referendum for declaration of war. Prior to the *Panay*, Representative Ludlow had accumulated only 205 of the 218 signatures needed to discharge his amendment from the Judiciary Committee, which had refused to refer it to the House. Within twenty-four hours after news of the *Panay* attack, enough additional signatures had been obtained to discharge it, and it came to a vote on January 10, 1938. Despite strong opposition from the President, the amendment was defeated in the House by a narrow margin of only 209 to 188, clearly indicating the residual strength of isolationism in Congress (Robert A. Divine, *The Illusion of Neutrality* [Chicago: University of Chicago Press, 1962], pp. 219–21).

[13]Lowe Chuan-hua, *Japan's Economic Offensive in China* (London: Allen and Unwin, 1939), p. 25.

however, that commercial Japanese vessels were being renamed with military designations and moving upriver with far more goods than required for army consumption. In fact, the Yangtze trade gradually became a Japanese monopoly, and it required steady American consular and diplomatic pressure in Shanghai and Tokyo to obtain permission for Stanvac to reestablish even minimum contact with its inland agents.[14] From late 1937 on, the company was never able to fully reestablish its trade in the Yangtze valley area.

Far to the north, the specter of a Japanese petroleum monopoly raised its head again, this time in Inner Mongolia, or Mengchiang, where in late 1937 the Kwantung Army had finally succeeded in establishing an "autonomous" government under its protege, Prince Teh.[15] On June 30, 1938, the managing director of the Manchuria Oil Company called on representatives of Stanvac, Asiatic Petroleum, and the Texas Company in Tientsin to announce that an exclusive oil marketing company would be formed in the Mengchiang area, with headquarters at Kalgan and capital subscribed at 450,000 yen by the Manchuria Oil Company, 200,000 yen by the Mengchiang Government Bank and Bus Company, and, it was hoped, 50,000 yen each by the three foreign oil companies. Would they participate? Stanvac's Tientsin office immediately contacted Yokohama, New York, and Ambassador Johnson

[14]Ibid., pp. 65–69; telegram, Frank P. Lockhart (Consul General at Shanghai) to SecState, May 11, 1928, *FRUS(1938)*, IV, 314–15; telegram, Allison (Third Secretary at Nanking) to SecState, June 9, 1938, *FRUS(1938)*, IV, 346–47; telegram, Allison to SecState, July 9, 1938, *FRUS(1938)*, IV, 400; telegram, Lockhart to SecState, August 11, 1938, *FRUS(1938)*, IV, 436–37; telegram, Smyth (Second Secretary at Nanking) to SecState, September 8, 1938, *FRUS(1938)*, IV, 466–68; Special Report No. S-17, U.S. Trade Commissioner in Shanghai, A. Viola Smith, November 4, 1938, Box 1178, "Peiping" folder, RG 151; letter, A. G. May (Stanvac, New York) to Maxwell M. Hamilton (Chief of the Division of Far Eastern Affairs), December 5, 1938, RG 59, 395.115 Standard-Vacuum Oil Company/78; letter, Ambassador Grew to Foreign Minister Matsuoka, September 18, 1940, *FRUS (Japan, 1931–41)*, I, 872–76.

[15]Crowley, *Japan's Quest for Autonomy*, pp. 306–308, 349.

in Hankow, while Asiatic alerted Shanghai, London, and Ambassador Craigie in Tokyo.[16] As in the Manchurian case, The Texas Company cooperated with the other two but did not engage in consultations with the State Department or Foreign Office. A further, odd, complication arose on July 12 when a representative of Toyoda Textile, Automotive, and Weaving Machines called at Stanvac's Tientsin office to enlist support for a covert effort on the part of Toyoda to replace the Manchurian Oil Company as controlling interest in the Mengchiang enterprise.[17] Stanvac, Asiatic, and The Texas Company would have no part of any of this proposition and requested diplomatic protests in Tokyo. As in Manchuria, little urging was required, and on July 21 and 22 Ambassadors Grew and Craigie called at the Japanese Foreign Office to protest the Mengchiang monopoly as a flagrant violation of the Open Door principle. Foreign Minister Ugaki professed little knowledge of the details and promised to investigate.[18]

A. G. May of Stanvac's New York office made clear to the Department of State that "the area itself was not important . . . [but] . . . it seemed likely that the purpose was to test the amount of opposition . . . [and] . . . if circumstances warranted the Japanese could extend the monopoly to other regions, eventually driving out the foreign oil companies altogether from

[16]Telegram, Johnson to SecState, July 2, 1938, *FRUS(1938)*, IV, 22–23; telegram, Asiatic Petroleum (Shanghai) to Asiatic Petroleum (London), July 1, 1938, copy in FO371(1938)F7313/295/10; letter, May to Hamilton, July 5, 1938, RG 59, 893.6363/163.

[17]Letter, P. J. Gallagher (Stanvac, New York) to Hamilton, July 14, 1938, RG 59, 893.6363/168; memorandum of July 27, 1938, prepared by Stanvac (Shanghai) for Lockhart, RG 59, 893.6363/190. The Toyoda representative is identified in the accessible record only as "Baron Ishimoto."

[18]Telegram, Hull to Salisbury (First Secretary at Peiping), July 7, 1938, *FRUS(1938)*, IV, 25; telegram, Foreign Office to Craigie, July 13, 1938, FO371 (1938) F7313/295/10; telegram, Grew to SecState, July 20, 1938, *FRUS(1938)*, IV, 30; telegram, Grew to SecState, July 21, 1938, *FRUS(1938)*, IV, 31; telegram, Grew to SecState, July 24, 1938, RG 59, 893.6363/174.

the Japanese-occupied portions of China."[19] Diplomatic haggling over the issue continued for another two months, when suddenly the Mengchiang government abandoned the whole plan and on September 24 abolished the oil monopoly. Grew learned privately in Tokyo that the scheme had been concocted by speculative businessmen who had assured the local government that the foreign oil companies would cooperate. When Stanvac, Asiatic, and The Texas Company stood their ground, support for the project collapsed, and in this case diplomatic defense of the Open Door succeeded.[20]

The British and American companies were thus blocked from resuming their Yangtze valley trade, but were clearly successful in opposing an Inner Mongolian monopoly. In the rest of North China they faced a more ambiguous situation. As early as April 1938, rumors of the formation of a Japanese-controlled North China Oil Company and the introduction of a rigid quota system had prompted formal words of caution by American, British, and Dutch representatives to the Japanese government in Tokyo.[21] Also in April, Asiatic Petroleum received overtures from Mitsubishi in London toward establishing some kind of cooperative marketing arrangement in North

[19]Memorandum of conversation, May with Ballantine (returned from China to an assignment in the Division of Far Eastern Affairs), August 2, 1938, RG 59, 893. 6363/178.

[20]Telegram, Grew to SecState, August 11, 1938, *FRUS(1938)*, IV, 38–39; letter, Grew to SecState, August 12, 1938, RG 59, 893.6363/189; telegram, Hull to Grew, August 26, 1938, *FRUS(1938)*, IV, 41–42; telegram, Grew to SecState, September 21, 1938, *FRUS(1938)*, IV, 46; telegram, Grew to SecState, September 29, 1938, *FRUS(1938)*, IV, 47–48; telegram, Grew to SecState, October 3, 1938, *FRUS(1938)*, IV, 53; Special Report No. S-17 by U.S. Trade Commissioner in Shanghai, A. Viola Smith, November 4, 1938, Box 1178, "Peiping" folder, RG 151; letter, May to Hamilton, November 7, 1938, RG 59, 893.6363/205.

[21]Telegram, Gauss (Consul General at Shanghai) to SecState, April 15, 1938, *FRUS(1938)*, IV, 13–14; Aide-Memoire from the American Embassy in Japan to the Japanese Ministry for Foreign Affairs, April 20, 1938, FO371(1938)F5284/295/10.

China, hinting that Stanvac had already made such a deal with Mitsui Bussan Kaisha in Shantung. Asiatic rejected the offer, and Parker emphatically denied to both Shell and the State Department that Stanvac had any arrangement with Mitsui other than its long-standing contract for bulk supplies destined for Japanese firms.[22] Some type of maneuvering for position was clearly afoot, however, and the next move came on June 30 when a Tientsin newspaper announced plans for a "North China Petrol Limited Company" capitalized eighty percent by the "Japan Petrol Company, Manchurian Petrol Company, Chosen Petrol Company, and Sun Petrol Company; 10 percent by Standard Oil, Texas Company, and Asiatic Petroleum; remaining 10 percent by owners of oil tankers."[23] Stanvac and Asiatic had heard nothing of this scheme, and by this time had made a firm agreement not to enter into any marketing schemes with Japanese capital.[24] Finally in mid-September the Japanese Minister of Finance and concurrently Minister of Commerce and Industry, Ikeda Seihin, informed Stanvac in Tokyo that plans for a North China Oil Company had been abandoned.[25]

[22]Letter, Mitsubishi Shoji Kaisha Ltd. (London) to Asiatic Petroleum (London), April 27, 1938, copy in FO371(1938)F5284/295/10; letter, O. W. Darch (Asiatic Petroleum, London) to Starling (Petroleum Department), May 2, 1938, copy in FO371(1938)F5284/295/10; letter, Parker to Hamilton, May 26, 1938, RG 59, 893.6363/159; telegram, Lockhart (Consul General at Shanghai) to SecState, May 27, 1938, RG 59, 893.6363/158.

[23]Telegram, Salisbury (First Secretary at Peiping) to SecState, July 2, 1938, *FRUS(1938)*, IV, 21–22. The company designations in this story are unclear; presumably, the "Japan Petrol Company" was Nippon Oil, and the "Manchurian" and "Chosen" Petrol Companies were the quasi-governmental organizations in those two areas. Essentially the same plan was described in Special Report No. S-17 by U.S. Trade Commissioner in Shanghai, A. Viola Smith, November 4, 1938, Box 1178, "Peiping" folder, RG 151.

[24]W. D. Allen (British Embassy, Shanghai) to Foreign Office, November 9, 1938, FO371(1938)F13328/62/10.

[25]Telegram, Grew to SecState, September 14, 1938, *FRUS(1938)*, IV, 44. Abandonment of the plan was confirmed in a private conver-

Petty local harassment continued for the next three years, but no actual Japanese monopoly of the North China petroleum trade ever materialized.[26] Stanvac, Asiatic, and The Texas Company had to endure difficulties in obtaining foreign exchange, problems with military permits for inland shipments, restrictive local quotas, price fixing designed to favor Japanese firms, and a host of other bureaucratic irritants. But from the

---

sation between C. E. Meyer of Stanvac, T. G. Ely of Rising Sun, Tani Masayuki (an official of the Japanese Foreign Office), and Nomura Shunkichi (a Japanese oil importer) in Tokyo on November 21, 1938; see memorandum of conversation in RG 59, 893.6363/206; and letter T. G. Ely (Rising Sun, Yokohama) to T. S. Powell (Asiatic Petroleum, Shanghai), November 26, 1938, copy in FO371(1939)F1498/534/10.

[26]Some contemporary accounts are misleading on this point, apparently because of continuing Japanese publicity about grandiose plans for the economic development of China. It is true that an impressive Asia Development Board was established under Cabinet jurisdiction on December 16, 1938, with the Premier as ex-officio president and the Ministers of Army, Navy, Foreign Affairs, and Finance as vice presidents. Among other functions this Board was to supervise the North China Development Company and the Central China Development Company, each designed to coordinate a series of other "companies," which would actually be consolidations of interests in specialized fields. It is also true that a "North China Oil Company" was carried on organization charts of the North China Development Company for some time, but in fact most of this grand scheme never materialized. The program bogged down for lack of complete military control over occupied areas, shortages of capital, indecision on ultimate policy, and lack of coordination at the working level. Only a few of the subsidiary companies were ever organized, and the North China Oil Company was not one of them. For a comprehensive summary of Japanese plans, see Lowe Chuan-hua, pp. 25–26, 51–53; telegram, Salisbury to SecState, April 25, 1938, FRUS(1938), III, 155–56; letter, Grew to SecState, December 23, 1938, FRUS(1938), III, 431–33. For a critique of the actual implementation, see memorandum of October 6, 1939, by Joseph M. Jones of the State Department's Far Eastern Division, FRUS(1939), III, 276–82. A useful summary of the political problems encountered in establishing the Asia Development Board (Kōa In) is given in Chitoshi Yanaga, Japan Since Perry (New York: McGraw-Hill, 1949), pp. 535–36.

fall of 1938 until late 1941 the three companies were able to carry on a substantial, though reduced, China trade.[27]

STANVAC AND SHELL thus adjusted to steady Japanese harassment of their marketing operations in both Japan and China, and with diplomatic support dug in to keep those territories reasonably open for a substantial though reduced volume of trade, but rapid changes in the political situation in East Asia and Europe forced Walden and Parker to recalibrate their thinking. Japanese troops appeared to be conquering China, capturing Hankow and Canton, and forcing Chiang to retreat to Chungking. Japan took a psychological step forward on November 3, 1938, when Prince Konoe reformulated the country's mission by proclamation of a "New Order in East Asia" to "stabilize" the area through the political, economic, and cultural cooperation of Japan, China, and Manchukuo—under Japanese leadership.[28] To Westerners, both rhetoric and map now suggested continuing Japanese expansion west and

[27]Special Report No. S-17 by U.S. Trade Commissioner at Shanghai, A. Viola Smith, November 4, 1938, Box 1178, "Peiping" folder, RG 151; letter, A. H. Elliot (Stanvac, New York) to Hamilton, February 15, 1940, RG59, 894.6363/340; letter, Grew to Japanese Minister for Foreign Affairs Matsuoka, September 18, 1940, *FRUS(Japan 1931–41)*, I, 872–76; letter, W. H. Pinckard (vice president, California Texas Oil Company, Ltd., New York) to Hornbeck, March 26, 1941, RG 59, 893.6363/244; telegram, Lockhart to SecState, June 26, 1941, RG 59, 393.115 Standard-Vacuum Oil Co./381. No reliable statistics could be located on the exact impact of Japanese restrictions, but the very fact that minor impediments far in the interior continued to be protested down to 1941 attests to continuation of the business. Physical damage to facilities of the three companies was not severe, and the kerosene trade required only periodic bulk shipments to inland distribution points. Available documents suggest that the companies also resorted to selling through Japanese middlemen in more instances than they cared to admit.

[28]Hugh Borton, *Japan's Modern Century* (New York: Ronald Press, 1955), p. 354; Butow, *Tojo*, pp. 120–21; and Jones, *Japan's New Order in East Asia*, pp. 78–79. An English translation of Konoe's statement is in *FRUS(1931–41)*, I, 477–81.

south. When Stanvac officers looked at *their* maps, the Japanese drive pointed directly toward the oil fields of the Indies.

Walden's personal sense of impending crisis had been considerably heightened on September 30, 1938, when the Western powers capitulated to Adolph Hitler at Munich, and thereafter he increasingly took command of Stanvac's relations with the United States government. The connection between Munich and Asia would appear tenuous until one considers Walden's background. As mentioned earlier, he had begun Jersey's original construction at Palembang and then had spent four years commuting between Batavia, The Hague, and New York to solidify organization of the Dutch-chartered NKPM.[29] As board chairman and chief executive officer of Stanvac after Walker's death in 1934, he had traveled extensively between New York, London, and The Hague, concentrating on production and legal problems and leaving much of the day-to-day marketing decisions in the hands of Parker. Until 1938 most of Stanvac's problems requiring diplomatic coordination had involved the Asian marketing network which Socony-Vacuum had brought to the merger, and Walden had let his "old China hand" president have a free rein. But to Walden, those were *his* oil fields in the Indies, and his global perspective caused him to feel increasing concern as the situation worsened in both Europe and Asia.[30] It was not a matter of conflict be-

[29]This description of Walden's background and viewpoint is based on a brief biographical sketch in *The Lamp*, February, 1934, pp. 20–27, Exxon library; an interview with Lloyd W. Elliott, managing director of NKPM in the late 1930s, in New York on August 25, 1970; and an interview with Edward F. Johnson, one of Stanvac's original vice presidents, in New York on August 11, 1971.

[30]Interview with Elliott. Readers who question this emphasis on the influence of one man's viewpoint on corporate policy might consider some additional data. Walden himself married late in life, left Stanvac at the outbreak of war to help coordinate petroleum movements for the American government in London, and after the war and the premature death of his wife, lived the rest of his life as a semi-recluse with his mother-in-law. His one really concrete accomplishment in life was the Palembang refinery, which was completely destroyed to

tween Walden and Parker, but a case where different circumstances called forth different leadership, and as chief executive officer Walden was well placed to exert control over governmental relations after 1938.

In any event, it happened that Walden was in The Hague at the height of the Czechoslovakian crisis, conferring with Lloyd W. ("Shorty") Elliott, then managing director of NKPM.[31] As Elliott recalled years later, he and Walden

---

prevent capture by the Japanese in 1942. All this would appear to support the hypothesis that Walden's real motivation in 1940 and 1941 was concern for *his* refinery, and he cast his lot with that institution (the American government) best able to resist the aggressor. He would not have been seriously opposed by other Stanvac officers, since protection of the Indies' investment would have been more important in 1940 and 1941 than the China market. And it is unlikely that he would have been opposed by Jersey or Socony-Vacuum management, who at that point were under heavy attack by the Temporary National Economic Committee for excessive concentration of power. The larger interests of the two parent corporations would have dictated cooperation rather than opposition to the American government. Although all these factors may have been operating, it is the author's opinion that Walden's personal motivation played the key role. Contrary to popular belief, it is not at all unusual for corporate managers to be dominated by other than a purely profit motive. For details on Stanvac's Asian investment, see Appendix A. On the parent companies' problems with the TNEC, see Nash, pp. 152–55; and U.S. Congress, Temporary National Economic Committee, *Investigation of Concentration of Economic Power, TNEC Monograph No. 39, Control of the Petroleum Industry by Major Oil Companies*, Senate Committee Print, 76th Cong., 3rd Sess. [1940] (Washington, D.C.: Government Printing Office, 1941); and *Investigation of Concentration of Economic Power: TNEC Monograph No. 39-A; Review and Criticism on Behalf of Standard Oil Company (New Jersey) and Sun Oil Company of Monograph No. 39 with Rejoinder by the Monograph Author*, Senate Committee Print, 76th Cong., 3rd Sess. [1941] (Washington, D.C.: Government Printing Office, 1941).

[31]Interview with Elliott, whose nickname Shorty derived from the fact that he was taller than almost all his associates; Elliott had also spent his career in the production side of the business and worked well with Walden.

listened to radio reports of Hitler's speech asserting that the Czech Sudetenland was the last of his territorial demands (September 26) and then talked late into the evening on the world situation. That night—in the atmosphere of imminent crisis which permeated European capitals at the time of Munich[32]—Walden and Elliott decided to begin planning what steps to take should the Japanese invade the Indies. Elliott returned to the Indies and set to work—first dismissing all German, Japanese, and Dutch employees of doubtful loyalty. Destruction plans for the refinery and oil wells began to be prepared openly, largely as a deterrent to Japanese action, and by the beginning of 1940 these were well advanced, as were plans for the early evacuation of most employees' families. In short, Walden and Elliott remained convinced from 1938 on that eventually Japan would attack the Indies. "It was just a question of when and how."[33]

This shift in corporate concern from active defense of marketing outlets to a closing of ranks against Japanese territorial expansion was not confined to the Standard-Vacuum Oil Company itself. In late December 1938, Frank A. Howard, president of the Standard Oil Development Company, another subsidiary of Jersey, came to the Department of State with a delicate problem.[34] The Japanese had now begun to show interest in the hydrogenation process they had rejected out of hand three years earlier. In this case, the Japanese Army was negotiating with I. G. Farben for assistance in building a plant in Shansi province in China, and the German corporation had

[32]Tension had mounted so high by the end of September that the British Navy had been put on the alert, and London began preparing for a possible air raid; Edward H. Carr, *International Relations Between the Two World Wars, 1919–1939* (New York: Macmillan, 1947), pp. 270–73.

[33]Interview with Elliott.

[34]Memorandum of conversation, Howard with Welles, Hornbeck, and Hamilton, December 28, 1938, RG 59, 893.6363/207; and letter, Godber (Asiatic Petroleum, London) to Starling (Petroleum Department), December 29, 1938, copy in FO371(1939)F1324/534/10.

approached the International Hydrogenation Company, jointly owned by Jersey and Shell, for a license to use its hydrogenation process—at an estimated fee of two million dollars. Howard talked with Undersecretary of State Sumner Welles, Hornbeck, whose title was now Political Advisor, and Maxwell Hamilton, who had replaced Hornbeck as Chief of the Far Eastern Division. What would the State Department advise? Hornbeck believed that Japan's chief motive was political—to demonstrate that American and British interests as well as German were aiding in Japan's industrial development of North China—and Welles suggested stalling as long as possible on the request. Shell made a similar inquiry at the Foreign Office, which checked with Washington and gave similar advice.[35] As a result, International Hydrogenation never gave permission for Japanese use of its hydrogenation process, and Japan did not acquire the data from Germany until late in World War II.[36] The incident indicated, however, that by 1938 the oil companies were looking to the State Department for direction on sensitive Asian issues.

Walden also made clear to his field personnel that should the United States decide to place an embargo on oil shipments to Japan, the company "would cooperate fully" and "stop all shipments from all properties under its control all over the world" even though much of that property was not under American jurisdiction. "Shipments from the Netherlands East Indies would be stopped despite the possibility that the Japanese Navy would attempt to take the properties there and despite the further fact that the American Government, due to cries within the United States against 'fighting for Standard Oil,' might not attempt to protect American interests in the

[35]Letter, Starling to Ronald (Foreign Office), February 4, 1939, FO371(1939)F1324/534/10; memorandum of conversation, F. R. Hoyer Millar (First Secretary of the British Embassy) with Hamilton, March 6, 1939, RG 59, 893.6363/208; telegram, Lindsay (British Ambassador in Washington) to Foreign Office, March 7, 1939, FO371(1939)F2297/534/10; and letter, Ronald to Starling, March 16, 1939, FO371(1939)F2297/534/10.

[36]Cohen, pp. 138–39.

Netherlands East Indies."[37] In short, by the beginning of 1939, Stanvac had dug in to hold whatever portion of the Chinese and Japanese market it could, begun preparations for the steps it would take should the Japanese attack the Indies, and made a firm decision to cooperate fully with the American government if the government chose to invoke economic sanctions.

FOR THE NEXT eighteen months, Stanvac remained in this stance of watchful waiting while the American government debated how it would respond to the rush of events. On economic sanctions, the issue was fully understood, but counsels divided. It was clear within the Department of State that any program of full sanctions must include an embargo on petroleum, and there was the possibility "that any attempt by the United States, Great Britain and the Netherlands to cut off from Japan exports of oil would be met by Japan's forcibly taking over the Netherlands East Indies."[38] But how *likely* was such a reaction? From Tokyo, Grew wrote that Japan had invested so much blood and effort on the mainland that sanctions alone were unlikely to cause her to abandon her "adventure in China." Instead he forecast a firming of resolution and would not favor "economic sanctions . . . unless the United States is prepared to resort to the ultimate measures of force." And American interests in Asia were not in Grew's opinion sufficient to warrant the risk of war.[39] Hornbeck viewed the issues differently on

[37]Letter, George A. Makinson (Consul General at Tokyo) to Grew, February 16, 1939, RG59, 894.24/868. This viewpoint was confirmed in the previously cited interviews with Elliott and Johnson and correlates well with subsequent events.

[38]Division of Far Eastern Affairs memorandum of December 5, 1938, RG59, 711.94/1234-1/2. This was a position paper prepared by Hamilton, Sayre, Hawkins, and Livesey, reviewed by Hornbeck, and forwarded to Welles. All but Hornbeck recommended a cautious policy.

[39]Letter, Grew to SecState, January 7, 1939, *FRUS(1939)*, III, 478–81. This conviction was deeply held and eloquently argued; Waldo H. Heinrichs, Jr., *American Ambassador: Joseph C. Grew and the Development of the United States Diplomatic Tradition* (Boston: Little, Brown, 1966), pp. 264–67.

both counts. He was now convinced that American rights and interests in East Asia required firm protection, and believed that sanctions would not create serious risk of armed conflict. Instead, such action "might contribute substantially toward obviating the development of a situation in which danger of armed conflict would become an actuality." In short, Hornbeck believed the Japanese would back down.[40] Views of others within the American diplomatic establishment ranged around and between these two poles, but public opinion in 1939 had begun to swing toward curtailment of supplies for Japan's war machine in China. Advocates of such an embargo tended to overlook the possibility that economic sanctions might lead to war, and Gallup polls in mid-1939 showed that two-thirds of those Americans questioned would join in a boycott of Japanese goods, while almost three-fourths favored an embargo on arms and ammunition to Japan.[41] Hull, with his usual cautious tenacity, preferred to stand pat. As Donald Drummond has put it, the Secretary of State was "much better at holding the line than in formulating new approaches."[42]

[40]Memorandum, Hornbeck to Sayre, December 22, 1938; RG59, 711.94/1234-1/2. Hornbeck persisted in this view, with increasing stridency, right down to December 1941; Thomas J. Farnham, "Stanley K. Hornbeck and the Coming of War with Japan, 1937–1941," paper presented at Duquesne University, October 27, 1971. Farnham provides a good synopsis of Hornbeck's views and points to the parallel between his "diplomatic war plan" of late 1937 and subsequent actions taken by the American government. In this author's opinion, however, events had more influence than Hornbeck himself on progressive American firmness in Asia.

[41]Langer and Gleason, *Challenge to Isolation*, p. 152. For a full discussion of this policy debate and the progressive toughening of American opinion, see ibid., pp. 147–159, 291–311; and Feis, pp. 17–48.

[42]Donald F. Drummond, "Cordell Hull, 1933–1944," in Norman A. Graebner, ed., *An Uncertain Tradition: American Secretaries of State in the Twentieth Century* (New York: McGraw-Hill, 1961), p. 193. For a further description of Hull's style, see Julius W. Pratt, *Cordell Hull, 1933–44*, in Robert H. Ferrell and Samuel F. Bemis, eds., *The American Secretaries of State and Their Diplomacy*, Vols. XII and XIII (2 vols.; New York: Cooper Square Publishers, 1964).

Circumstances, however, did not permit a completely static position. Hull had already invoked a "moral embargo" in June 1938, by requesting that American firms refrain from shipping aircraft or aerial armament to Japan since these might be used to attack civilians. By mid-1939, however, pressure had mounted for sterner measures. In April, Key Pittman, Chairman of the Senate Foreign Relations Committee, introduced a resolution authorizing the President to place an embargo on export-import trade with any party to the Nine-Power Treaty found to be threatening the lives or legal rights of American citizens—obviously Japan.[43] Partly to avoid having that resolution came to a vote, Hull threw his support behind a resolution sponsored by Senator Arthur Vandenburg, calling for six-months notice of termination of the 1911 Treaty of Commerce and Navigation between the United States and Japan. When this resolution appeared headed for a bitter floor fight, Hull and Roosevelt decided to act unilaterally, and gave notice of termination to the Japanese Ambassador on July 26. The action, which had no legal impact other than *permitting* restrictive American action after January 26, 1940, caught the Japanese government by surprise, generating considerable concern over actual American intentions. The issue was eclipsed almost immediately, however, by announcement of the Soviet-German Non-Aggression Pact on August 22, and the outbreak of war in Poland on September 1.[44]

Worry mounted in Tokyo as Poland collapsed, the "phony war" settled over Europe, the "China Incident" remained unresolved, and January 26 arrived with no clear indication of American economic intentions.[45] Grew later referred to this

[43]This account of the termination of the 1911 Treaty of Commerce and Navigation is based on Feis, pp. 21–24; and Langer and Gleason, *Challenge to Isolation*, pp. 157–59. The text of the termination notice is in *FRUS(1931–41)*, I, 189; and the text of the treaty itself is in *FRUS(1911)*, pp. 315–19.

[44]Basic agreement was reached August 22; the pact was dated August 23, but not actually signed until August 24; Langer and Gleason, *Challenge to Isolation*, pp. 181–83, 187.

[45]Lu, pp. 62–67.

posture as a deliberate "sword of Damocles"[46]—an apt analogy except that it provided no way to predict how the nervous party might behave. In Japan, policy makers increasingly eyed the Netherlands East Indies as a source of supply should America act, and on February 2, 1940, Japan notified The Hague that it wished to open negotiations on expanded trade with the Indies.[47] The Dutch displayed no haste in replying, and immediately after Hitler's invasion of Denmark and Norway on April 9, Foreign Minister Arita issued a press statement warning that Japan could not "but be deeply concerned over any development . . . in Europe that might affect the *status quo* of the Netherlands East Indies . . . " because Japan was "economically bound by an intimate relationship of mutuality" with that colony.[48] Hull immediately issued a counterstatement, recalling that in 1922 the United States, Britain, France, and Japan had all formally resolved to "respect the rights of the Netherlands in relation to their insular possessions in the region of the Pacific Ocean."[49] No direct threat, but a clear word of caution. Events moved rapidly thereafter: On May 10 Germany invaded the Netherlands; on May 20 Arita presented a renewed request for negotiations to the Dutch Ambassador in Tokyo; on June 6 the Dutch government in London gave a polite but noncommittal reply; and on June 22 France capit-

[46]Diary entry for July, 1940, Vol. 101, p. 4438, Grew MSS.

[47]Hubertus J. van Mook, *The Netherlands Indies and Japan: Battle on Paper, 1940–1941* (New York: Norton, 1944), pp. 26–29. This account, by the senior Dutch participant in the Indies negotiations in 1940 and 1941, is especially useful for its reproduction of a number of otherwise inaccessible documents. See also Jan O. M. Broek, *Economic Development of the Netherlands Indies* (New York: Institute of Pacific Relations, 1942), pp. 116–18.

[48]Press Release by the Japanese Embassy on April 15, 1940; *FRUS (Japan, 1931–41)*, II, 281.

[49]Department of State Press Release of April 17, 1940; *FRUS(Japan, 1931–41)*, II, 281–82. The assurance, issued in the form of four identical notes to the Netherlands government, had been in lieu of honoring a Dutch request to join in the Four-Power Treaty at the Washington Conference; Buckley, p. 141.

ulated to Adolf Hitler.[50] Events in Europe had suddenly changed the power structure in Southeast Asia, and Japan wanted to negotiate in earnest—especially on oil.

ALL DURING the spring of 1940 Stanvac personnel in general and Walden in particular had kept in close touch with the American government on oil movements in East Asia. Considerable data on Japanese stockpiling and Indies refinery capacity were provided to the Bureau of Foreign and Domestic Commerce, the Office of Naval Intelligence, and the Department of State,[51] and the company began to check regularly with both the Navy and the State Department before quoting on Japanese orders for questionable items such as blending agents, which could be used for producing aviation gasoline.[52] Concurrently with the fall of France and increased Japanese pressure on the Netherlands East Indies, the informal coordination between Stanvac and the Department of State which had developed in the early 1930s was resurrected to deal with a new and far more serious problem. The threat of a Japanese move against the oil fields of the East Indies, so lightly brushed aside in 1934, now appeared a real possibility.

[50]Van Mook, pp. 29–43; Broek, pp. 116–18.

[51]Letter, Frank A. Howard (president, Standard Oil Development Company) to Joseph C. Green (Chief, Division of Controls), March 18, 1940, RG59, 894.24/867; memorandum, Harold W. Moseley (Division of Controls) to Green, April 23, 1940, RG59, 894.24/906; letter, Paul L. Hopper (Fuels Section, Metals and Minerals Division, Bureau of Foreign and Domestic Commerce) to Walden, March 25, 1940, Box 1660, Folder: "Oils-Mineral-Japan, 1926–1940," RG 151; memorandum, F. L. Correll to Director, War Plans Division, April 9, 1940, file (SC) JJ7-6/ff6, Records of the Office of the Chief of Naval Operations, Division of Naval History, Navy Yard, Washington, D.C., hereafter cited as "CNO Records"; memorandum, Moseley to Charles W. Yost, April 11, 1940, RG59, 856D.6363/755.

[52]Memorandum, Yost to Green, March 26, 1940, RG59, 894.24/964; memorandum of conversation, Hornbeck with Capt. R. E. Schuirmann (Office of the Chief of Naval Operations), June 7 and 8, 1940, RG59, 894.24/961; letter, Walden to Rear Admiral W. S. Anderson (Director of Naval Intelligence), June 10, 1940, file (SC) JJ7-3/EE37, CNO Records.

FIVE

HULL PROTECTS THE INDIES

IN THE early summer of 1940 Asian specialists in general and Stanvac management in particular became convinced that some type of Japanese move against the Indies was imminent—especially if the United States responded to the growing clamor for an embargo on oil.[1] Although their concern derived partly from rumor and partly from the logic of the situation, it was by no means groundless. Unknown to Westerners at the time, pressure for just such action was building within the Japanese government as the war in Europe created drastic shifts in the world balance of power. In mid-July, the government of Prime Minister Yonai Mitsumasa fell because of army impatience that it was not moving rapidly enough toward full alignment with the spectacularly successful forces of Hitler and Mussolini,[2] and responsibility for forming a new cabinet passed to Prince Konoe Fumimaro. Even before he had completed the selection of his cabinet, Konoe called three key ministers-designate to his home in Ogikubo on July 19 for a discussion of basic policy. Out of this "Ogikubo Conference" came a set of guidelines adopted by the full Cabinet on July 26, ratified by a Liaison Conference with officers of the Supreme Com-

---

[1]Telegram, Grew (Tokyo) to SecState, May 14, 1940, *FRUS(1940)*, IV, 18; telegram, Johnson (Chungking) to SecState, June 4, 1940, *FRUS(1940)*, IV, 23; memorandum by Hamilton, June 4, 1940, *FRUS (1940)*, IV, 576; telegram, Patton (Consul General at Singapore) to SecState, June 22, 1940, *FRUS(1940)*, IV, 32; circular telegram, British Undersecretary of State for Dominion Affairs to Commonwealth High Commissioners, July 23, 1940, copy in FO371(1940)F3634/677/23; memorandum of conversation, Henry F. Grady (Assistant Secretary of State), James C. Dunn (Advisor on Political Relations), and Hamilton with George Walden, July 25, 1940, *FRUS(1940)*, IV, 55–56.

[2]Butow, *Tojo*, pp. 138–40; Feis, pp. 76–78.

mand the following day, and used as a blueprint for Japanese planning for months thereafter.[3] The guidelines included stronger phraseology on Japan's "Southern Policy" than any previous statement of similar authority. They said in part:

In order to include the English, French, Dutch and Portuguese settlements within the substance of the New Order in the Far East, positive arrangements will be undertaken . . . . Even though we will avoid unnecessary collisions with the United States, as long as it concerns the establishment of a New Order in the Far East, we are firmly determined to eliminate any interference by show of power.[4]

[3]Butow, *Tojo,* pp. 142–43, 147–153; Feis, pp. 84–87. Butow *(Tojo,* p. 153) suggests that "the decisions of July 26 and 27, 1940, constituted a pivotal stage in the metamorphosis of the China Incident into the Greater East Asia War" and argues that the policies were foisted on gullible civilians by aggressive staff officers of the military bureaucracy. However, Crowley has argued persuasively that—as in the case of Manchuria and China—there may have been differences over tactics, but Konoe and other key civilians fully subscribed to the evolving sense of imperial mission (Crowley, "A New Deal for Japan and Asia," p. 258). During the late 1930s and especially in 1941, the Liaison Conference became the key decision-making body within both the Japanese government and the military high command, which technically exercised power independently under separate authority from the Emperor. The meeting on July 27 was the first such conference in two years, but they were held with increasing frequency thereafter; Butow, pp. 149–50; Nobutaka Ike, trans. and ed., *Japan's Decision for War: Records of the 1941 Policy Conferences* (Stanford: Stanford University Press, 1967), pp. xv-xix. For a full discussion of this decision-making process, see Maxon, *passim.*

[4]Harada, *Saionji-Harada Memoirs,* chapter 373, 19 August 1940. This is Harada's version of the original decisions of the Ogikubo Conference. The final version adopted by the Liaison Conference, entitled "Gist of the Main Points in Regard to Dealing with the Situation to Meet the Change in World Conditions," was somewhat briefer and softer. It included a specific plan to "strengthen the diplomatic policy towards the DUTCH EAST INDIES in order to obtain important materials"; Exhibit No. 1310, International Military Trubunal, Far East, *Record of Proceedings* (Tokyo, 1946–48), copy in the National Archives, Washington, D.C., hereafter cited as "IMTFE."

Not quite a decision to seize the Indies by force if necessary, but inching toward it.

In practical terms, this policy reinforced steps already taken by the Japanese government to acquire a larger share of Indies oil, and Walden had already begun to brace himself to deal with the pressure. On July 16 the Japanese Foreign Office had informed Dutch Minister General Pabst in Tokyo that the government wished to send a trade delegation to Batavia for "economic negotiations," and on July 26 the Netherlands government reluctantly agreed.[5] By this time, rumors of an American oil embargo had prompted representatives of the Japanese government to open negotiations in Yokohama with Rising Sun for a share of Indies production over and above the quota normally allocated to the two companies, and Stanvac had agreed in advance to provide its pro rata share (about twenty-seven percent) of whatever quantity was decided upon.[6] As rumors of an American embargo grew stronger and Japanese concern mounted, oil became the primary item on the proposed Batavia agenda, and the Yokohama negotiations were transferred to Batavia for consummation.[7] Walden and

[5]Van Mook, p. 45; Feis, p. 96; memorandum of conversation, R. A. B[utler] (British Undersecretary of State for Foreign Affairs) with Dr. C. J. I. M. Welter (Dutch Minister for the Colonies), July 25, 1940, FO371(1940)F3634/677/23; Walter A. Foote (American Consul General at Batavia) to SecState, November 13, 1940, RG 59, 756D.94/103.

[6]Lloyd Elliott (Stanvac, Soengi Gerong, Sumatra) to Walden, July 19, 1940, and telegram, Arthur Elliot (Stanvac, London) to Walden July 22, 1940, both in RG 59, 856D.6363/756-1/9; memorandum of conversation, Grady, Dunn, and Hamilton with Walden, July 25, 1940, FRUS(1940), IV, 55–56; letter, F. Godber (Shell, London) to Sir Cecil Kirsch (Petroleum Department) August 18, 1940, and letter, F. C. Starling (Petroleum Department) to C. E. Steel (Foreign Office), August 26, 1940, both in FO371(1940)W9435/9160/49.

[7]Telegram, Grew to SecState, August 29, 1940, FRUS(1940), IV, 90; telegram, Foote to SecState, September 14, 1940, FRUS(1940), IV, 115. Formal transfer of the Yokohama negotiations to Batavia did not take place until late September (telegram, Kobayashi [Chief of the Japanese Mission at Batavia] to Matsuoka, September 18, 1940, Exhibit 1315, IMTFE; telegram, Foote to SecState, September 26, 1940, FRUS(1940), IV, 154–55).

Parker together with their counterparts in Shell decided for psychological reasons to reinforce local management in the Indies with representatives from the head offices, and while the Japanese were still trying to fix the composition of their own delegation, Stanvac vice president Fred H. Kay and Shell's J. C. van Panthaleon Baron van Eck were sent out to coordinate negotiations in Batavia.[8] Before tracing the course of these negotiations, however, it might be well to note simultaneous developments in the United States.

AT THE SAME time that the Japanese government was beginning to think seriously of moving south to mitigate the effects of a possible oil embargo, efforts had begun in the United States to invoke just such an embargo to check Japanese expansion. Although Stanvac was not the moving force behind attempts to cut off Japan's oil, Walden was deeply involved on the periphery of this effort and was fully aware of the divisions of opinion within the American government. Legal authority for action had been created rather innocuously on July 2 with passage of a National Defense Act designed in part to streamline procurement of war materials. Section VI of that act permitted presidential control of critical exports, and Roosevelt used his military authority as Commander in Chief to designate Lieutenant Colonel Russell L. Maxwell as Administrator of Export Control. There were two key restrictions. Maxwell's independent agency established guidelines and procedures, but actual export licenses were issued by the Department of State's Division of Controls, and Maxwell was instructed confidentially to defer to Hull on all questions affecting foreign policy. As long as the matter involved bona fide procurement problems, Maxwell reigned supreme, but as soon as the machinery began to be used for economic warfare the State Department

[8]Van Mook, p. 48; memorandum of conversation, Hornbeck and Ballantine with Walden, Kay, Wilkinson (Shell, New York), and Maris (British Embassy, Washington), August 16, 1940, *FRUS(1940)*, IV, 75; letter, Walden to Hornbeck, August 20, 1940, RG 59, 756.94/173.

held a double administrative veto.[9] In practice, a selective embargo could be enforced by issuing a presidential proclamation requiring licensing of *all* exports of a given commodity and then granting licenses for that commodity freely except to selected countries, such as Japan.

This jury-rig arrangement, which lasted almost a year, received its first acid test within three weeks of its establishment, when Treasury Secretary Morgenthau tried to use it to override the Department of State and impose a full embargo on petroleum for Japan. Most accounts of Morgenthau's attempted coup tend to overlook or minimize the fact that a genuine interest had aready developed within the Department of State for checking what appeared to be a sudden Japanese attempt to corner the American market in high-grade aviation gasoline[10]—an interest intensified by information coming to Hornbeck from his chief contact in the oil business, Walden. Reports of suddenly increased Japanese purchasing had begun to reach the

[9]Feis, pp. 72–75; a well annotated collection of directives issued by the Administration of Export Control prior to World War II is contained in *Documents on American Foreign Relations*, Vol. III, July 1940-June 1941, ed. by S. Shepard Jones and Denys P. Myers (Boston: World Peace Foundation, 1941), pp. 473–98; and *Documents on American Foreign Relations*, Vol. IV, July 1941-June 1942, ed. by Leland M. Goodrich (Boston: World Peace Foundation, 1942), pp. 718–40. An "inside" account of how the agency functioned is contained in an undated typescript manuscript by Roberta R. Johns, "An Historical Record of Export Control, Book One: The Office of the Administrator of Export Control—A Military Function, July 2, 1940–Sept. 15, 1941," Box 887, Item 145, Record Group 169, Records of the Foreign Economic Administration, National Archives Branch, Suitland, Maryland, hereafter cited as "RG 169."

[10]See, for example, Langer and Gleason, *Challenge to Isolation*, pp. 720–22; and Blum, II, 348–54. Feis is an exception; in *The Road to Pearl Harbor* (pp. 88–89) he clearly points to prior State Department concern over aviation gasoline purchases. His account, however, does not mention the direct role that Stanvac played in this episode, and he does not appear to have used the records of Maxwell's office, which were found to be interfiled with those of the wartime Foreign Economic Administration (RG169).

Department by July 17,[11] and Walden made an urgent trip to Washington on July 18 to inform Hornbeck that "during the last 48 hours" Japanese firms had placed orders in the United States for immediate delivery of somewhere between 300,000 and 400,000 barrels of high-grade aviation gasoline, with more requests for quotes still coming in. Although he could not deduce whether this unusual surge of activity foretold preparations for military operations, Walden was seriously worried. He indicated he would be glad to forego Stanvac participation in these sales if State wished to impose some type of general embargo.[12] Hornbeck immediately discussed the matter with Political Advisor James C. Dunn and Chief of Naval Operations Admiral Harold R. Stark, and on the 19th penned a strong recommendation to Welles that prompt action be taken to block excessive shipments of aviation gasoline to Japan.[13] Before Welles could act, however, he was called into a meeting with Roosevelt to discuss *another* proposal on oil originating from an altogether different source.

On the preceding evening (the 18th) Secretary of the Treasury Morgenthau had dined with British Ambassador Lord Lothian, newly appointed Secretary of War Henry Stimson, Secretary of the Navy Frank Knox, the Australian Minister, and several other British officials. When Stimson needled Lo-

[11]Caltex board chairman James A. Moffett had talked with Hamilton on the 17th regarding Japanese purchases of gasoline for delivery in Thailand, and Joseph C. Green, Chief of the Controls Division, had discussed this briefly with Maxwell. Then on the 18th, Frank A. Howard, president of the Standard Oil Development Company, had reported increased Japanese purchasing of aviation gasoline in the course of a conversation with Harold W. Moseley of the Division of Controls on the subject of tetraethyl lead (memorandum, Green to Hamilton, July 17, 1940, RG 59, 894.24/1036; and memorandum, Moseley to Green, July 18, 1940, RG 59, 894.24/1040).

[12]Memorandum of conversation, Hornbeck with Walden, L. V. Collings (Stanvac) and C. A. Thompson (Stanvac), July 13, 1940, RG 59, 811.20(D) Regulations/163.

[13]Memorandum, Hornbeck to Welles, July 19, 1940, RG 59, 811.20 (D) Regulations/164; part of this memo is printed in *FRUS(1940)*, IV, 586–87.

thian on the recent British acquiescence to Japan in temporarily closing the Burma Road, Lothian retorted that the United States for similar reasons had not curtailed the sale of aviation gasoline to Japan. As the discussion grew more animated, a grandiose idea emerged which Morgenthau enthusiastically presented to Roosevelt the following morning.[14] Japan and Germany might be brought to heel if the United States could stop *all* oil exports on grounds of national defense requirements, Britain could obtain its supplies from the Caribbean, a joint Anglo-American effort could be made to buy up world surpluses, Allied bombing could be concentrated on synthetic oil production in Germany, and the Dutch could be persuaded to destroy the oil wells in the Indies.[15] Whatever this plan may have lacked in technical feasibility, it made up in global imagination. Roosevelt showed sufficient interest to call in and try it out on Stimson, Knox, and Welles (substituting for Hull, who had left that day for a meeting in Havana). Although well aware of a specific problem developing on aviation gasoline, Welles was aghast at so sweeping a provocation toward Japan, and the meeting ended inconclusively.[16] By coincidence, this was the afternoon of July 19, the same date as Konoe's Ogikubo Conference in Japan.

Records of what occurred during the next week present a confusing picture, but a careful reconstruction strongly suggests that Roosevelt and Welles made a firm decision on July 22 to block only excessive exports of aviation gasoline and held to that policy despite Morgenthau's energetic attempts to

[14]Diary entry for July 18, 1940, Henry L. Stimson MSS, Sterling Library, Yale University, New Haven, Connecticut, hereafter cited as "Stimson MSS"; memorandum of conversation, Morgenthau with White and Klotz, July 19, 1940, Morgenthau Diary, Vol. 284, pp. 201–11. The dinner and the subsequent conference with Roosevelt are summarized in Feis, pp. 89–90; Langer and Gleason, *Challenge to Isolation*, p. 721; and Blum, II, 348–51.

[15]Memorandum for the President, July 19, 1940, Morgenthau Diary, Vol. 284, p. 122.

[16]Diary entry for July 19, 1940, Stimson MSS; memorandum of conversation, Morgenthau with Foley, White, and Klotz, July 19, 1940, Morgenthau Diary, Vol. 284, pp. 212–16.

generate a full embargo.[17] Morgenthau may have speeded up action on aviation gasoline, however, because he did send a telegram to Roosevelt at Hyde Park on July 22 warning that "valuable strategic material [was] slipping through our fingers every day."[18] That same day Roosevelt telephoned Welles and directed that "measures [be] taken with the least possible delay to control the exportation of aviation gasoline and lubricating oil for aircraft engines, in order to conserve these materials in the interest of national defense."[19] Joseph C. Green, Chief of the Division of Controls, promptly drafted the necessary documents and sent them to Maxwell for processing through other interested agencies. As it turned out, State's original definition of "aviation gasoline" was modified by request of the National Defense Advisory Commission and the Army-Navy Munitions Board.[20] At the request of the Treasury Department, the term "petroleum products" was substituted at one point in the text for "aviation motor fuel" and "aviation lubricating oil," but the full set of documents approved by Roosevelt on July 26 clearly applied controls only to aviation products.[21] If the

[17]This interpretation differs from that of Feis (pp. 92–93), Langer and Gleason, *Challenge to Isolation* (p. 722), and Blum (II, 353) who believed that Roosevelt was vacillating. It is based on documents from the office of the Administrator of Export Control apparently not utilized heretofore and a careful rereading of records of the Cabinet debate on July 26.

[18]Telegram, Morgenthau to Roosevelt, July 22, 1940, Morgenthau Diary, Vol. 285, p. 1.

[19]Letter, Green to Maxwell, July 22, 1940, Item 171, Tab D1, Box 994, RG 169; another copy of this letter is in RG 59, 811.20(D) Regulations/152D.

[20]Memorandum, W. A. Harriman (National Defense Advisory Commission) and Col. Charles Hines (Army-Navy Munitions Board) to Maxwell, July 25, 1940, Item 97, Box 225, RG 169.

[21]Memorandum, Herbert E. Gaston (Treasury Department) to Maxwell, July 25, 1940, Item 171, Tab D1, Box 994, RG 169; Johns Manuscript, pp. 26–28, Item 145; Box 887, RG169. The final proclamation and explanatory regulations are printed in *FRUS(Japan, 1931–41)*, II, 216–18. These regulations also dealt with aviation lubricating oil, tetraethyl lead, and iron and steel scrap, but in the interest of brevity discussion of these less critical items has not been included in this study.

rhetorical thunder from the Treasury could be disregarded, this action appeared to stand on its own merit as a reasonable response to a legitimate problem. Walden reported to the State Department that week that Japanese orders for immediate delivery of aviation gasoline had reached 972,000 barrels, ranging between 87 and 92 octane,[22] and the Army's Fourth Corps Area commander in San Francisco warned Washington on the 23rd that the "market will be stripped . . . [and] . . . should Army and Navy need it in quantity during next six to nine months there would be [a serious] shortage of aviation gasoline."[23]

Considerable confusion and press speculation arose, however, from a flank attack mounted by Morgenthau. Independent of State's action, Morgenthau himself drafted a proclamation placing controls on all "petroleum and petroleum products" without the qualifying limitation to aviation gasoline and sent this directly to Roosevelt while the other documents were still being circulated for approval by other agencies.[24] Presumably thinking this method of setting up the procedure had been cleared with State—as all such questions were supposed to be—Roosevelt signed Morgenthau's version and sent it to the State Department for affixing the Great Seal before promulgation. Thanks to this administrative routine, Welles stopped the risky version and made sure the President signed the correct one.[25] At a Cabinet meeting on the 26th, however, Morgenthau attacked Welles again. Although the under-

[22]Memorandum of conversation, Grady, Dunn, and Hamilton with Walden, July 25, 1940, RG 59, 894.24/1035.

[23]Radiogram, Dewitt (San Francisco) to Adjutant General, July 23, 1940, Morgenthau Diary, Vol. 286, p. 234. At Morgenthau's request, Stimson sent copies of this radiogram to Morgenthau, Ickes, and Roosevelt (diary entry for July 24, 1940, Stimson MSS).

[24]Letter, Morgenthau to Roosevelt, July 22, 1940, Morgenthau Diary, Vol. 285, p. 185; a copy of Morgenthau's draft is in RG 59, 811.20(D) Regulations/324.

[25]Feis, p. 92; transcript of telephone conversations, Morgenthau with Grady, Stimson, and Ickes, all on July 26, 1940, Morgenthau Diary, Vol. 287, pp. 151, 154–61.

lying issue was undoubtedly establishing the legal framework for either a full or a partial embargo on oil to Japan, the open argument dealt with whether or not too much paper work would be generated if *all* petroleum exports were subjected to licensing.[26] It appears to have been this administrative bickering, not the substantive question, that caused Roosevelt to throw up his hands, refuse to participate, and tell the two men to "go off in a corner and settle their issue."[27] Nowhere in the available documentation is there evidence that Roosevelt seriously considered anything other than a check on excessive exports of high-octane aviation gasoline to Japan in the summer of 1940. Be that as it may, the restrictions on aviation gasoline that went into effect on July 26 had the psychological effect of intensifying Japanese concern over the possibility of being cut off altogether from American supplies and simply heightened Japanese pressure on the Indies. As had been the case in Japan, Manchuria, and China, Stanvac found itself caught in the middle—but this time the stakes were considerably higher.

As IT TURNED OUT, the hastily drafted definition of aviation gasoline issued on July 26 almost gave Morgenthau what he wanted anyway, and this administrative detail became the hidden crux of American oil policy for the next twelve months. The regulations of July 26 specified that henceforth licenses would be required for the export of all:

[26]This was the thrust of a memorandum that Morgenthau took with him to the Cabinet meeting, and it was this administrative challenge to which Welles later gave a formal reply. Morgenthau did not tell his staff the *substance* of the debate at the Cabinet meeting, and Stimson only noted in his diary that there was a "grand discussion" between the two; memorandum addressed to the President but noted as having been seen only by Welles, July 26, 1940, Morgenthau Diary, Vol. 287, pp. 162–64; letter, Welles to Morgenthau, July 29, 1940, Morgenthau Diary, Vol. 288, pp. 43–55; transcript of staff meeting, July 26, 1940, Morgenthau Diary, Vol. 287, pp. 128–9; diary entry for July 26, 1940, Stimson MSS.

[27]Diary entry for July 26, 1940, Stimson MSS.

Aviation motor fuel, i.e., high octane gasolines, hydro-carbons, and hydrocarbon mixtures (*including crude oils*) boiling between 75° and 350° F. which with the addition of tetraethyl lead up to a total content of 3 c. c. per gallon will exceed 87 octane number by the A.S.T.M. Knock Test Method; *or any material from which by commercial distillation there can be separated more than 3% of such gasoline, hydrocarbon or hydrocarbon mixture.*[28] (Italics added).

State's Division of Controls and the Office of the Administration of Export Control were immediately deluged with requests for clarification of governmental intent because—as Jersey representatives told the Division of Controls on August 2—this definition could be interpreted to require export licenses for "approximately 90% of all crude oils produced in the United States . . . substantially all motor gasolines . . . [and] . . . all the light products of petroleum up to kerosene."[29] To avoid this administrative nightmare, Maxwell's office obtained the consent of the Army-Navy Munitions Board and the National Defense Advisory Commission to drop the "crude oil" and "3%" clauses, and on August 6 limited export licensing to:

[28]Regulations . . . [issued] . . . July 26, 1940, *FRUS(Japan, 1931–41),* II, 217–18. The original draft prepared by the Department of State on July 22 had read:

Aviation motor fuel—Gasolines, iso-octane, neo-hexane, alkylates, and other petroleum components having an octane rating of 87 or higher, including any fuel from which a constituent having an octane rating of 87 or higher can be obtained by distillation.

(Memorandum, Green to Maxwell, Item 171, Tab D1, Box 994, RG 169.) The working papers show that this was modified several times before arriving at the one promulgated on July 26.

[29]Memorandum by Charles W. Yost (Division of Controls) reporting a meeting with Orville Harden, Henry Bedford, Henry Dodge, and D. Jennings of Standard Oil (New Jersey), August 2, 1940, RG 59, 811.20(D) Regulations/255.

high octane gasoline with an octane number by the A.S.T.M. Knock Test Method of 87, or more; or gasoline which with the addition of tetraethyl lead up to total content of 3 c. c. per gallon, or less, will have an octane number by the A.S.T.M. Knock Test Method of 87 or more; or synthetic iso-octane, isopentane, neohexane or alkylates; or mixtures which are the result of mixing synthetic iso-octane, isopentane, neohexane, or alkylates with other materials, including crude oils.[30]

After three weeks of rumor and confusion, the working level of the American government, influenced by a cautious Department of State, had settled on a narrow definition that would permit conservation of the specific types of high-octane gasoline threatened by excessive Japanese purchasing without inundating the government in paper work on peripheral items. From a strategic viewpoint, however, this unpublicized administrative action simply forced the Japanese to purchase lower octane—but perfectly usable—aviation gasoline, and as long as this definition remained in effect, there would be no serious blockage to Japanese procurement of petroleum sup-

[30]Directive No. 58 of August 6, 1940, Item 177, folder: "Export Policy Licensing Memoranda, 1940–42," Box 1021, RG169. The more comprehensive definition of July 26 printed in the *Foreign Relations* series was in effect only ten days. The revision issued on August 6 and not printed in *Foreign Relations* remained in effect until June 27, 1941; Directive No. 427, June 27, 1941, Item 177, folder: "Export Policy Licensing Memoranda, 1940–42," Box 1021, RG 169; and memorandum by Major J. S. Bates on "Control of Export of Petroleum Products," June 1941, folder: "History of Export Control Exhibits, 1940–41, Tab D1," Box 994, RG 169. Morgenthau heard that "there were leaks being made" in the embargo by the Munitions Board, and at his request Stimson called in Maxwell and two members of the Board for interrogation. They assured Stimson that while the embargo was narrow, it had effectively stopped the threatened run on aviation gasoline (Diary entries for August 6 and 7, 1940, Stimson MSS; and transcript of telephone conversation, Morgenthau and Stimson, August 7, 1940, Morgenthau Diary, Vol. 290, pp. 197–200).

plies from the United States. A grasp of this point is critical to an understanding of much of what followed.[31]

WHILE THIS quiet adjustment was being made at the working level, thunder and lightening continued to emanate from the pinnacle of the Treasury Department, and Stanvac found itself involved here also. Beginning on July 29, Treasury's Division of Research and Statistics took on the unlikely chore of distributing weekly reports on petroleum exports to the President and anyone else who might be interested, carefully annotating them to show how much oil was being shipped to Japan in excess of the *original* regulations of July 26.[32] Then on August 2, acting on his own initiative, Morgenthau asked the British government to "send over at once and secretly, a leading British oil man to discuss the whole situation."[33] Lothian cabled London, requesting "someone who has a real practical knowledge of the oil business and who will get on well both with American oil companies and the State Department and Treasury."[34] Apparently Morgenthau sought an ally and failed to

[31]While this was being resolved, the Japanese Ambassador repeatedly expressed concern over the possible hostile *intent* of the new regulations; memorandum of conversations with the Japanese Ambassador by Sumner Welles on July 26, July 27, August 3, and August 23, 1940, *FRUS(1940)*, IV, 589–91, 595–6, 598–9.

[32]These weekly reports are scattered all through official files of the period. A copy of the first is attached to a letter from Assistant Secretary Herbert Gaston to the President, July 29, 1940; "Treasury" folder, Box 37, President's Secretary's File, Franklin D. Roosevelt MSS, Franklin D. Roosevelt Library, Hyde Park, New York, hereafter cited as "Roosevelt MSS." The annotation used to imply violation of a Presidential directive became increasingly sophisticated as time went on; see, for example, the eight-month summary attached to a letter from Morgenthau to Ickes, September 17, 1941, Item 4, Box 474, File 710, Record Group 31, Records of the Petroleum Administration for War, National Archives, Washington, D.C., hereafter cited as "RG 31."

[33]Telegram, Lothian to Foreign Office, August 2, 1940, FO371 (1940)W9160/9160/49.

[34]Ibid.

realize that Lothian in his dinner conversation of July 18 had spoken entirely on his own initiative, without a shred of authority from his government. When the original grandiose scheme had been cabled to London for comment, the Foreign Office had been dismayed. Not only did the scheme threaten disruption of Britain's oil supply, but it risked provoking a Japanese attack in Southeast Asia with no assurance of American support.[35] For the next three months, while Hitler's effort peaked and ebbed, the Foreign Office worked to discourage rash American action on Asian oil, and its response to Morgenthau's request for someone to discuss petroleum was made in this context. The task was given to Agnew of Royal Dutch-Shell, well known to both Stanvac and the State Department, and then serving as chairman of Britain's wartime Petroleum Board. Agnew was thoroughly briefed on the tenuous situation in Southeast Asia and sailed for the United States in mid-August.[36] By coincidence, Agnew arrived in the United States just as the Batavia negotiations came to a head and was well situated to help integrate the cautious positions of Stanvac,

[35]Telegram, Lothian to Foreign Office, July 19, 1940; Foreign Office minutes of July 20 through July 27, 1940; memorandum, Churchill to Lord Ismay, July 21, 1940; all in FO371(1940)F3634/677/23. The Foreign Office records on this episode are well summarized in William N. Medlicott, *The Economic Blockade*, in *History of the Second World War, United Kingdom Civil Series* (2 vols; London: H. M. S. O., 1952–59), I, 476–81.

[36]Telegram, Foreign Office to Lothian, August 6, 1940, FO371 (1940)W9160/9160/49; letter, Agnew to Lord Halifax (Foreign Secretary) August 9, 1940, FO371(1940)W9433/9160/49; telegram, Foreign Office to Lothian, August 14, 1940, FO371(1940)W9433/9160/49; and minutes of meeting at the Petroleum Department, August 14, 1940, FO371(1940)W9547/9160/49. The last recorded a session with representatives of the Foreign Office, Ministry of Economic Warfare, Air Ministry, Petroleum Department, and Petroleum Board, three days before Agnew sailed. It was agreed that Agnew should cooperate as fully as possible with the American government but try to deflect any further provocation on oil, since such moves tended to direct the "force of Japanese demands into quarters which might make difficulties for [Britain]."

Shell, the State Department, and the Foreign Office into a consistent policy. From a tactical standpoint, Morgenthau's action misfired.

Once Morgenthau tackled a project, however, his enthusiasm was boundless. On August 7 he arranged a full briefing on world petroleum for himself and two of his presumed allies in the Cabinet, calling in officers of Jersey and Stanvac to discuss the Dutch East Indies, Japanese reserves, tetraethyl lead, the octane count required for military flying, and a number of related topics.[37] Present at the meeting were Morgenthau, Secretary of the Interior Ickes, Secretary of the Navy Knox, Jersey vice presidents E. J. Sadler and Frank Howard, Stanvac board chairman Walden, and several supporting experts. At Morgenthau's request, Walden opened the discussion with a detailed review of Asia, providing his hosts with annotated maps and extensive statistics on both Japan and the Indies. At one point, the Treasury Secretary interrupted to point out that the United States had embargoed the shipment of aviation gasoline, and asked, "If we wanted to put on a few more screws to make it more difficult for Japan to become aggressive, in what direction would be the next move?" Without hesitation, Walden replied, "I think, Mr. Secretary, that if pressure is so great as to impoverish them of oil in Japan, they will move south to the Dutch East Indies."[38] While directed to Morgenthau, this comment appeared to have more impact on Knox,

[37]A verbatim transcript of this meeting is in the Morgenthau Diary, Vol. 290, pp. 46–95, with additional papers in Vol. 292, pp. 260–74. Morgenthau, Ickes, and Knox held a similar, but less productive meeting with William F. Humphrey, president of Tidewater Associated Oil Company the next day (Morgenthau Diary, Vol. 290, pp. 204–17), and Sadler, Howard, and Humphrey all sent additional data to Morgenthau after the meetings; letters to Morgenthau from E. J. Sadler, August 9, 10, and 12, 1940, and letter from Howard to Morgenthau, August 13, 1940, Morgenthau Diary, Vol. 294, pp. 92–104; letter, Humphrey to Morgenthau, September 11, 1940, Morgenthau Diary, Vol. 306, pp. 185–91. A summary of this meeting also appears in Blum, II, 355; and Ickes, III, 297–99.

[38]Morgenthau Diary, Vol. 290, p. 55.

who apparently had just begun to realize the connection. Knox began to waver, staying behind after the meeting to tell Morgenthau that he would have to "think about our Fleet and where we will be willing to send it."[39] The Treasury had gained a wealth of information but frightened an ally.

Knox's concern may have derived from intensive strategic planning underway in the Navy and War Departments in the summer and fall of 1940. From mid-1939 on the Joint Board had been debating which of five "RAINBOW" War Plans to adopt for full development since each assumed a different strategic situation, and the war in Europe was changing the balance of forces with alarming speed. The fall of France in June 1940 forced a grim reassessment, and by November the Joint Board had agreed with a recommendation of Admiral Stark that the United States adopt a "Europe first" strategy and proceed posthaste with completion of the War Plan RAINBOW 5.[40] This plan, fully known to both Hull and Roosevelt, as-

[39]Memorandum of August 7, 1940, Morgenthau Diary, Vol. 290, p. 27.

[40]This summary of the status of RAINBOW 5 in the fall of 1940 is based on Louis Morton, *Strategy and Command: The First Two Years*, in the series *United States Army in World War II: The War in the Pacific* (Washington, D.C.: Department of the Army, 1962), pp. 21–91. Morton provides an interesting account of the evolution of American military strategy in the Pacific, especially the long disparity between ambitious diplomatic objectives and limited military means surrounding work on War Plan ORANGE from 1907 to 1938, assuming war with Japan only. So-called RAINBOW Plans began to be developed in 1939, when it appeared possible that the United States could be at war with different combinations of foreign powers. The decision to expand RAINBOW 5 in November of 1940 led to development of a joint Anglo-American plan, ABC-1, in the spring of 1941, which became the actual basis of Allied strategy in World War II. An earlier version of Morton's account is contained in his "Germany First: The Basic Concept of Allied Strategy in World War II," in Kent R. Greenfield, ed., *Command Decisions* (Washington, D.C.: Department of the Army, 1960), pp. 11–47. An expanded account of joint Anglo-American military planning in the spring of 1941 is contained in Maurice Matloff and Edwin M. Snell, *Strategic Planning for Coalition Warfare, 1941–1942*, in the series *United States Army in World War*

sumed that the United States would act in concert with Britain toward the early defeat of Germany and Italy, maintaining a "strategic defensive . . . in the Pacific until success against European Axis Powers permitted transfer of major forces to the Pacific for an offensive against Japan." While this decision, which became the basis for American strategy in World War II, had not fully matured in the fall of 1940, the discussions leading to it were well underway at the time of the Batavia negotiations. Hull knew, therefore, that Admiral Stark and General George Marshall preferred to avoid a conflict with Japan for as long as possible, and Walden's actions suggest that he was well aware of this trend in top level American governmental thinking.

To recapitulate, Stanvac entered the Batavia negotiations not as a free agent, but as one party to an intricate network of bureaucratic, political, and strategic influences tending toward some type of temporary accommodation with Japan. Fortunately, there was no fundamental disagreement among the five parties most directly involved. Walden for two years had foreseen the possibility of a Japanese attack on the Indies. With neither Britain nor the United States ready or able to block such a move (especially considering the precarious status of the British Empire in the fall of 1940), Walden considered it temporarily prudent to accede to Japanese demands, distasteful though such accommodation might be. Shell's Agnew was of a similar mind for similar reasons, and both companies were being pressed by the Dutch government-in-exile in London to avoid antagonizing Japan. Regardless of its normally aggressive stance, the Foreign Office also recognized the risk of widening hostilities during the Battle of Britain and actively sought an accommodation. In the Department of State, Hull, with the backing of Roosevelt,[41] likewise hoped to minimize the risk, and had the support of Welles and most of the Far

---

II: The War Department (Washington, D.C.: Department of the Army, 1953).

[41]Despite the debate within his Cabinet, Roosevelt himself never

Eastern Division in this posture. Most critical in the fall of 1940, however, was the administrative check that State held on the actual machinery of export control. Exponents of a full embargo—such as Morgenthau, Stimson, Ickes, and Horn-beck—simply found themselves outside the line of authority and unable to influence the actual course of events. The Japanese had no way of knowing all this, however, and carefully watched both the rhetoric and the actual mechanics of export controls before deciding how hard to press in Batavia.

THE MOST critical aspect of the negotiations began offstage, unknown to the general public, and continued quietly, parallel with the actual talks. Almost immediately after the August 6 switch in the technical definition of aviation gasoline, specialists close to the problem realized that although this might protect American supplies of very high-octane aviation gasoline, it had almost no effect on Japan's ability to purchase lower octane but perfectly satisfactory aviation fuel in the American market. Most Japanese aircraft operated quite effectively at 80 to 87 octane, and purchasing simply shifted to gasoline just below the 87-octane limitation. In actual practice, this middle-octane aviation gasoline continued to be shipped from the west coast in quantities limited only by the number of tankers avail-

---

appears to have wavered on this point in the fall of 1940. In a private conversation on oil and scrap iron with Roosevelt on August 16, Morgenthau found that the President "talked in the same vein as S. Welles, namely we must not push Japan too much at this time as we might push her to take Dutch East Indies" (Morgenthau "Presidential Diary," Vol. 3, p. 644). The subject of partial embargoes against Japan came up in a meeting with Cordell Hull, Sumner Welles, and Breckinridge Long on October 10, and "the President's position was that we were not to shut off oil from Japan . . . and thereby force her into a military expedition against the Dutch East Indies but we were to withhold from Japan only such things as we vitally needed ourselves, such as high test gas" (Fred L. Israel, ed., *The War Diary of Breckinridge Long: Selections from the Years 1939–1944* [Lincoln: University of Nebraska Press, 1966], p. 140.).

able to carry it.[42] Thanks to Morgenthau's weekly statistics this situation became an open secret within the American government, but problems arose when various administrators assumed that the intent had been to deny *all* aviation gasoline to Japan and tried to plug what they construed as a loophole in the system. In each case they ran afoul of State's absolute veto on changes affecting American foreign policy.

First in the field was the National Defense Advisory Commission, which on August 14 unsuccessfully suggested restoring a revised version of the "3% clause."[43] Next came the Director of Naval Intelligence, Rear Admiral Walter S. Anderson, who reported these apparent violations to Knox in late August with the suggestion that Jersey's Sadler be consulted on development of "airtight" specifications. At Knox's request, Anderson forwarded his memoranda to Morgenthau where they died for lack of access to the true line of authority.[44] In

[42]Telegram, Lothian to Foreign Office, August 31, 1940, FO371 (1940)W9996/9160/49; memorandum by Green (Division of Controls), September 7, 1940, RG59, 811.20(D) Regulations/1798; memorandum of September 26, 1940, Morgenthau Diary, Vol. 309, p. 12; letter, Walden to Hornbeck, October 24, 1940, RG59, 894.6363/363; memorandum, A. H. McCollum to Director of Naval Intelligence, November 2, 1940, file (SC) L11-4/EF37, CNO Records; letter, Walden to Sadler, November 4, 1940, copy in RG59, 894.24/1159; and memorandum by Maj. J. S. Bates on "Control of Export of Petroleum Products," June 1941, folder "History of Export Control Exhibits, 1940–41," Tab D1, Box 994, RG169. Feis acknowledges in a footnote that this situation existed (p. 99, n. 7), but attributed it to "faulty specifications." The specifications were not faulty; they accomplished exactly what had been originally intended—no more and no less. Medlicott (I, 482–83) notes that the British were fully aware of this situation and declined to press the issue since they opposed the embargo in the first place.

[43]Letter, W. A. Harriman (National Defense Advisory Commission) to Maxwell, August 14, 1940, Item 97, "Administrator of Export Controls, Central Files," Box 225, RG169.

[44]Memorandum, Anderson to CNO, August 26, 1940, file (SC) JJ7, CNO Records; memorandum, Anderson to SecNav, August 30, 1940, file (SC) L11–4/EF37; letter, Anderson to Morgenthau, August 29, 1940, Morgenthau Diary, Vol. 296, pp. 196–97; letter, Anderson

September an informal coalition of technical experts from the office of the Administrator of Export Control, the National Defense Commission, and State's Division of Controls attempted to resolve the problem, enlisting the aid of Stanvac to develop a satisfactory working definition.[45] Walden and Sadler consulted Jersey's experts in Baton Rouge and provided a set of possible specifications, but when all this came to the attention of State's Far Eastern Division, Hamilton killed the project on the grounds that with negotiations "actively in progress in Batavia . . . it would not from a political point of view be advisable to widen . . . the scope of existing regulations."[46] Undaunted by these rebuffs, Maxwell himself took up the cause on October 2, suggesting reinstatement of the original regulations of July 26. In his best dilatory style, Hull "considered" Maxwell's repeated requests for a decision for almost three months, and finally on December 27 sent word that "in the opinion of the Department, considerations of foreign policy make it advisable that . . . [a broadened defi-

---

to Morgenthau, August 30, 1940, Morgenthau Diary, Vol. 296, pp. 215–16; letter, Anderson to Morgenthau, September 17, 1940, Morgenthau Diary, Vol. 318, pp. 312–13. This correspondence was used by James H. Herzog to argue that Anderson worked at cross purposes with Stark; see his "Influence of the United States Navy in the Embargo of Oil to Japan, 1940–1941," *Pacific Historical Review*, XXXV, No. 3 (August 1966), pp. 317–28. Although Anderson may have believed in firmness toward Japan, the correspondence only suggests a naval officer wishing to see regulations enforced as he understood them to be written—not someone making a serious effort to undercut Stark's "Europe first" position.

[45]Memorandum by Green, September 7, 1940, RG59, 811.20(D) Regulations/1798; report of meeting, Maj. J. S. Bates (Administration of Export Control) with Sadler, Walden *et al.*, September 9, 1940; and letter, Sadler to Bates, September 27, 1940, both in Item 97, "Administrator of Export Controls, Central File," Box 226, RG169; memorandum by Green, September 17, 1940, RG59, 811.20(D), Regulations/904.

[46]Memorandum, Hamilton (Chief, Far Eastern Division) to Controls Division, September 23, 1940, RG59, 811.20(D) Regulations/905.

nition] . . . be not issued at the present time."[47] Through these quiet administrative vetoes, Hamilton and Hull effectively ensured that the "embargo" of July 26 would have no significant effect on Japan's ability to procure American petroleum, and this action immeasurably eased pressure on the Indies in the fall of 1940.

WHILE ALL this discussion had been transpiring in the United States, Stanvac and Shell had attempted to develop a realistic strategy to counter Japanese demands in Batavia. The original note handed the Dutch Ambassador in Tokyo on May 20 had requested one million tons of "mineral oil" annually from the Indies, but as nervousness over a possible embargo increased, that figure was raised to two million tons on June 29.[48] As already noted, negotiations for increased shipments were

[47]Memorandum, Green to Maxwell, December 27, 1940, RG59, 811.20(D) Regulations/943. Maxwell's repeated attempts to get a decision are covered in: memorandum, Maxwell to Green, October 2, 1940, RG59, 811.20(D) Regulations/942; memorandum, Price (Division of Controls) to Yost, November 14, 1940, RG59, 811.20(D) Regulations/902; memorandum, on "Exports to Japan" prepared by Yost, November 15, 1940, RG59, 811.20(D) Regulations/1151; memorandum, Maxwell to Green, December 11, 1940, RG59, 811.20(D) Regulations/960; and assorted interdepartmental notes attached to the final memorandum to Maxwell on December 27, 1940, RG59, 811.20 (D) Regulations/943.

[48]Van Mook, pp. 31–32; telegram, Grew to SecState, July 1, 1940, *FRUS(1940)*, IV, 39–40; memorandum of conversation, Butler (Foreign Office) with Welter (Dutch Minister for Colonies), July 25, 1940, FO371(1940)F3634/677/23; Medlicott, I, 480. It is impossible to convert from tons to barrels without knowing the specific gravity of the products involved and whether the reference is to metric tons (2204.62 lbs.), British long tons (2,240 lbs.), or short tons (2,000 lbs.). As a rule of thumb, however, one metric ton equals approximately seven U.S. 42-gallon barrels. The request for 2,000,000 tons annually would therefore have equalled approximately 14,000,000 barrels, compared with the 4,824,000 gallons which Stanvac and Shell shipped to Japan from the Indies and Borneo in 1939 (Appendix B, Table B-6). The term "tons" is used throughout this account of the Batavia negotiations because this was the term used almost exclusively in the primary documents.

opened with Rising Sun in Yokohama by representatives of the Japanese Navy and the Planning Board (speaking for the Army and a group of refiners), and by July 25 Walden had decided that it would be prudent for Stanvac to comply with a Dutch request that his company provide its pro rata share (about twenty-seven percent) of whatever quantity was agreed upon.[49] Before confirming this, however, Walden went down to Washington to ensure approval of the State Department. In a conference with Assistant Secretary of State Henry Grady, Political Advisor James Dunn, and Hamilton, he was told that the Department "realized the difficulties of the situation confronting" him and only cautioned him to avoid "making it possible for Japan to obtain unusually large quantities of petroleum products" and if possible to enter into contracts "for a limited period of time."[50] With this tacit approval, Walden proceeded to work out the best arrangement possible.

By August 16 Stanvac and Shell had agreed to go as far as they could in meeting Japanese demands and to send Kay and van Eck to the Indies to coordinate the actual negotiations. Walden and Shell's representative in the United States cleared this posture with the British Embassy in Washington and then met with Hornbeck for his reaction.[51] It was unexpectedly violent! According to the British report of the meeting, Hornbeck "made it clear that he disliked concessions to the Japanese in general and this one in particular since . . . it included . . . aviation gasoline and very high octane crudes." He would not object to low grade crudes, but aviation stocks smacked of

[49]Letter, F. Godber (Shell) to Sir Cecil Kirsch (Petroleum Department), August 18, 1940, copy in FO371(1940)W9435/9160/49; memorandum of conversation, Grady, Dunn, and Hamilton with Walden, July 25, 1940, *FRUS(1940)*, IV, 55–56.

[50]Memorandum of conversation, Grady, Dunn, and Hamilton, July 25, 1940, *FRUS(1940)*, IV, 55–56.

[51]Telegram, Godber (Shell, London) to Wilkinson (Shell, New York), August 13, 1940, FO371(1940)W9435/9160/49; memorandum of conversation, Hornbeck and Ballantine with Walden, Kay (Stanvac), Maris (British Embassy), and Wilkinson (Shell, New York), August 16, 1940, *FRUS(1940)*, IV, 75–79.

"appeasement," and in his opinion "resistance of outrageous demands was more likely to maintain peace in the Pacific than concessions."[52] The outburst caused all parties to regroup but was probably Hornbeck's last influential statement on the Batavia negotiations. The Foreign Office, concluding that Hornbeck was "pigheaded" and out of touch with reality and his colleagues, decided to try to deal directly with Welles or Hull.[53] Hull himself had begun to realize the dangers of Hornbeck's violently anti-Japanese viewpoint and had decided to circumvent him by talking directly with Hamilton on sensitive matters.[54] It was Walden, however, who completed the neutralization of Hornbeck on the Batavia talks. Knowing his man, Walden returned on August 22 to obtain Hornbeck's approval on a draft cable to London stating that Stanvac itself would provide crude, but had no aviation gasoline available since its entire output in the Indies was already contracted to the British government. Walden also agreed to tell Shell that Stanvac would cooperate only because requested to do so by the Dutch government, and not because it was "in sympathy with the policy involved." Since Shell was willing to supply some aviation gasoline, this did not affect the substance of the

[52]Telegram, Lothian to Foreign Office, August 17, 1940, FO371 (1940)W9554/9160/49. Hornbeck's memorandum of the meeting, cited in the preceding footnote, presents the same thoughts in more temperate language. A third report of the same meeting is attached to a letter, Kirsch (Petroleum Department) to Steel (Foreign Office), September 5, 1940, FO371(1940)W10142/9160/49. A copy of this same report, presumably written by Maris, and dated August 16, 1940, is in Box 67, "Petroleum" folder, Stanley E. Hornbeck MSS, Hoover Institution on War, Revolution and Peace, Stanford University, Stanford, California, hereafter cited as "Hornbeck MSS."

[53]Telegram, Lothian to Foreign Office, August 17, 1940, with attached Foreign Office minutes of August 21, 1940, FO371(1940) W9554/9160/49.

[54]Israel, p. 139. Hull considered Hornbeck "almost alone in advocating measures that almost certainly would bring . . . [the United States] . . . into war" (Moffat Journal, Vol. 46, "Notes on visit to Washington, October 6th to 10th [1940]," p. 1, Jay Pierrepont Moffat MSS, Houghton Library, Harvard University, Cambridge, Massachusetts.)

proposed transaction, but it satisfied Hornbeck's idea of the United States standing on principle and he agreed to offer "no objection."[55] For the next three months Walden sent Hornbeck copies of all key cables passed between New York, London, and Batavia and date stamps on the documents reveal that they were promptly routed to Hamilton's Far Eastern Division.[56] Hornbeck, Hamilton, and Hull thus had direct access to information on every move in the Batavia negotiations, and placed no further obstacles in the way of Walden's attempts to reach a minimum-risk settlement.

Despite strong interest in the Indies, the Japanese were slow to complete the organization of their delegation to Batavia. Three names were advanced to lead the party and then were withdrawn. Finally, the government settled on Kobayashi Ichizo, Minister of Commerce and Industry, who reached Batavia on September 12 heading a group of thirty-nine diplomatic, military, and economic experts.[57] Included in that delegation were thirteen petroleum experts who arrived a few days ahead of Kobayashi under the leadership of Mukai Tadaharu, now board chairman of Mitsui Bussan Kaisha, and an old acquaintance of Stanvac's Yokohama office.[58] Originally, the

[55]Memorandum of conversation, Hornbeck with Walden and Prioleau (Stanvac), August 22, 1940, RG 59, 856D.6363/763.

[56]Letters, Walden to Hornbeck, September 16, 18, 26, 27, October 9, 11, 24, 25, 31, November 8 and 13, 1940, all in RG 59, 856D. 6363/756 to 798. See also telegram, Hull to Foote (Consul General in Batavia), September 11, 1940, *FRUS(1940)*, IV, 108–109.

[57]Van Mook, pp. 46–48; telegram, Grew to SecState, August 28, 1940, *FRUS(1940)*, IV, 88–89. A useful overview of the Indies negotiations is contained in Muhammed A. Aziz, *Japan's Colonialism and Indonesia* (The Hague: Martinus Nijhoff, 1955), pp. 121–40. For a reliable interpretation of reports on the Batavia talks as reflected in State Department records, see Feis, pp. 96–100, 104, 123–24, 130–32. For a similar view of the talks as reflected in the records of the Foreign Office, see Medlicott, I, 481–84, and II, 76–80.

[58]The number of oil experts in the advance party is given as eight by Van Mook (p. 48) and in a telegram, Foote to SecState, September 5, 1940, *FRUS(1940)*, IV, 96; but the names and positions of thirteen individuals are listed in a letter from Walden to Hornbeck, September 11, 1940, RG 59, 856D. 6363/757-3/9.

Japanese had wanted to negotiate with Governor General Jonkheer Tjarda van Starkenborgh Stachouwer, but since the discussions were billed as economic rather than political the Dutch insisted that they deal with Director of Economic Affairs Hubertus J. van Mook, elevated to the rank of minister plenipotentiary for the occasion.[59] Elliott, managing director of Stanvac's NKPM in Palembang found himself asked to serve as one of the technical experts backing up van Mook; he had worked so closely with the Dutch that they considered him one of their own.[60] As previously mentioned, Kay and van Eck also arrived to represent the New York and London offices of Stanvac and Shell.

Kobayashi attempted to steer the initial discussions toward a general agreement on political and economic cooperation, but it quickly became apparent that petroleum ranked as the major concern. When the Dutch refused to discuss a Japanese request for oil-producing concessions, Kobayashi turned to increased shipments of crude and refined products, presenting a demand for 3,150,000 tons annually under a five-year contract guaranteed by the Indies government. This was three times the figure advanced in May and six times the quantity scheduled to be shipped in 1940 under the two companies' normal sales quota! Declining to serve as a "broker," van Mook suggested that the Japanese deal with Stanvac and Shell on a commercial basis.[61] Accordingly, on September 25, the

[59]Van Mook, pp. 47–48; telegram, Grew to SecState, August 28, 1940, *FRUS(1940)*, IV, 88–89; telegram, Craigie to Foreign Office, FO371(1940)W10556/9160/49; telegram, Foote to SecState, September 14, 1940, *FRUS(1940)*, IV, 115–16.

[60]Interview with Elliott in New York on March 4, 1970.

[61]Van Mook, pp. 49–54; telegrams, British Consul General at Batavia to Foreign Office, September 13 and 20, 1940, FO371(1940) W10556/9160/49; letter, Walden to Hornbeck, September 18, 1940, RG 59, 856D.6363/756-5/9; telegram, Foote to SecState, September 19, 1940, RG 59, 756D.94/69; telegram, British Consul General at Batavia to Foreign Office, September 26, 1940, FO371(1940)W10679/ 9160/49; memorandum of conversation, Butler (Foreign Office) with Welter (Dutch Colonial Minister), September 27, 1940, FO371(1940)

Yokohama negotiations were formally transferred to Batavia, and Mukai was assigned responsibility for concluding an agreement directly with Stanvac and Shell.[62] He promptly reiterated the demand for 3,000,000 tons, and Kay and van Eck had no way to determine whether this was in earnest or simply a bargaining ploy.[63]

In the meantime, Agnew had arrived in the United States and begun to work with Stanvac on the details of a joint position. After conferring with Walden in New York, Agnew went down to Washington to meet with Morgenthau on September 3, Hornbeck on September 4, and Hull on September 5, in all cases accompanied by Lothian. After these meetings he sent a long memorandum to Morgenthau and Hornbeck.[64] Obviously attempting to minimize obstacles to a negotiated settlement, Agnew argued that cooperation between the United States and Britain was absolutely vital on oil policy, and in the present instance he hoped that the American government would "understand, even if it does not approve, the commitments which

---

W10697/9160/49; telegram, British Consul General in Batavia to Foreign Office, October 4, 1940, FO371(1940)W10901/9160/49; letter, Foote to SecState, November 13, 1940, RG59, 756D.94/103. Stanvac and Shell's combined sales quota in Japan for 1940 was 494,000 tons (letter, Walden to Hornbeck, September 27, 1940, RG 59, 856D.6363/777).

[62]Telegram, Foote to SecState, September 26, 1940, *FRUS(1940)*, IV, 154–55.

[63]Letter, Walden to Hornbeck, September 26, 1940, RG 59, 756D. 94/92.

[64]Telegram, Lothian to Foreign Office, August 26, 1940, FO371 (1940)W9854/9160/49; memorandum of conversation, Morgenthau and White with Lothian and Agnew, September 3, 1940, Morgenthau Diary, Vol. 302, pp. 152–3; memorandum of conversation, Hull and Lothian, Agnew, and Casey (Australian Minister), September 5, 1940, *FRUS(1940)*, IV, 97; telegram, Lothian to Foreign Office, September 5, 1940, FO371(1940)W10170/9160/49; memorandum, "The Far East—Oil," September 12, 1940, Morgenthau Diary, Vol. 306, pp. 172–76; letter, Agnew to Hornbeck, September 17, 1940, RG 59, 894.24/1087-1/2.

the Dutch and . . . [British] . . . may be forced to make."[65] Whether this round of conciliatory talks had any significant effect is problematic, but it may have helped to hold the more bellicose forces at bay during a critical period. As soon as the Japanese demand for 3,000,000 tons reached New York, Walden and Agnew went to work on what turned out to be an astute response. Since the sales were to be purely "commercial" they calculated the maximum efficient capacity of their present Indies facilities, deducted outstanding commitments, and arrived at a combined Stanvac/Shell total of 2,000,500 tons which could be made available. They reduced this at van Mook's request to 1,849,500 tons for bargaining purposes (including the preexisting combined quota of 494,000 tons), allocated shares so that Stanvac would not be shipping aviation gasoline, offered the crude on twelve-month contract (to satisfy agreements already made in Yokohama), and the finished products on six-month contract (to satisfy the American request for short-term commitments). Finally, and most significantly, they decided that all but the original quota business should be f.o.b. the Indies so that the Japanese would have to provide the tankers. By cable between New York, Washington, London, and Batavia all details were cleared with the Foreign Office and the Indies government, and on October 3 Walden and Agnew reviewed the proposed offer with Hornbeck, who again offered no objection, but urged that the final settlement be kept within the limits specified.[66]

This was no ordinary commercial transactions in ordinary times. On September 23, Japanese troops began to occupy

[65]Memorandum of September 12 attached to letter, Agnew to Hornbeck, September 17, 1940, RG 59, 894.24/1087-1/2.

[66]Letter, Walden to Hornbeck, September 26, 1940, RG 59, 756D. 94/92; telegram, Lothian to Foreign Office, September 27, 1940, FO371(1940)W10738/9160/49; telegram, Lothian to Foreign Office, October 3, 1940, FO371(1940)W10842/9160/49; telegram, Foreign Office to Lothian, October 8, 1940, FO371(1940)W10842/9160/49; letter, Walden to Hornbeck, October 9, 1940, RG 59, 856D. 6363/772. Agnew's account of these negotiations is attached to a letter, Kirsch (Petroleum Department) to Seymour (Foreign Office), November 11, 1940, FO371(1940)W11898/9160/49.

northern Indochina, and on September 27 news broke that Japan had signed a Tripartite Alliance with Germany and Italy, promising to aid one another if "attacked by a Power at present not involved in the European War or in the Chinese-Japanese conflict."[67] All this precipitated a heated Cabinet debate over Asian policy on October 4, with Roosevelt holding firm to his "Europe first" position.[68] On October 10 he reiterated to Hull and Welles that "we were not to shut off oil from Japan . . . and thereby force her into a military expedition against the Dutch East Indies."[69] During this debate Roosevelt also told Morgenthau to get out of the oil business and let the State Department take care of "handling foreign affairs."[70] The times were tense, and the consensus hammered out by Stanvac, Shell, the Foreign Office, and the Department of State over the Batavia negotiations proved to be in complete concert with Roosevelt's Asian policy in the fall of 1940.

On October 8, van Mook transmitted to Mukai a formal offer by Kay and van Eck of 1,849,500 tons of crude and products, distributed as shown in Table V-1 (see page 154).[71] Mukai consulted Tokyo for a decision, and then on October 19 completely surprised the Dutch by accepting the offer just as it stood, with a subdued comment that additional quantities might be requested in the future.[72] He then entered into detailed

---

[67]Text of the Tripartite Pact of September 27, 1940, League of Nations, *Treaty Series* (205 vol.; Geneva, 1920–46), CCIV. 386.

[68]Diary entry for October 4, 1940, Stimson MSS.

[69]Israel, p. 131.

[70]Memorandum of meeting, October 2, 1940, Morgenthau Diary, Vol. 318, pp. 121–27. Morgenthau grudgingly complied. His Diary vividly reflects where his time was spent, and it is filled with documents on oil from mid-July through the first of October 1940, where this documentation abruptly ends.

[71]Van Mook, pp. 65–72; letter, Foote to SecState, November 13, 1940, RG 59, 756D.94/103.

[72]Letter, Walden to Hornbeck, October 11, 1940, RG 59, 856D. 6363/756–6/9; telegram, Foote to SecState, October 21, 1940, RG 59, 856D.6363/778; telegram, British Consul General at Batavia to Foreign Office, October 22, 1940, FO371(1940)W10901/9160/49; letter, Foote to SecState, November 13, 1940, RG 59, 756D.94/103.

TABLE V — 1

STANVAC/SHELL OFFER TO JAPANESE ECONOMIC
DELEGATION IN BATAVIA, OCTOBER 8, 1940
(In long tons of 2,240 lbs.)

Products	Quantities Requested	Quantities Offered
Aviation crude (from Shell)	1,100,000	120,000
Other crude oils	1,150,000	640,000
Aviation gasoline (over 87 octane)	400,000	33,000[a]
Gasoline (under 71 octane)	Nil	250,000
Other products	500,000	312,500
Subtotal	3,150,000	1,355,500
Preexisting import quotas	Nil	494,000
Total crude and products	3,150,000	1,849,500

[a]Under prior contract with Shell, but not yet delivered.

SOURCE: Van Mook, pp. 65–72; and letter, Foote to SecState, November 13, 1940, RG 59, 756D.94/103.

discussions with three Indies oil men, one of whom was Elliott, on terms and specifications.[73] Within three weeks detailed agreements had been worked out including all of the terms specified by Walden and Agnew plus a requirement that payment be made in American dollars, and these were signed by Mukai, Kay, and van Eck on November 12.[74] Technically the Batavia talks moved on to other matters, but Kobayashi sailed for Japan on October 22, just after agreement on oil had been reached in principle, and a lull settled over the negotiations

[73]Interview with Elliott, March 4, 1970.

[74]Telegram, British Consul General at Batavia to Foreign Office, November 13, 1940, FO371(1940)W10901/9160/49; telegram, Foote to SecState, November 13, 1940, FRUS(1940), IV, 207–8; letter, Foote to SecState, November 13, 1940, RG 59, 756D.6363/103. Complete texts of the final agreements with both Stanvac and Shell are in RG 59, 856D.6363/800.

until a new delegation chief arrived in late December. Hull's administrative veto over any significant curtailment of American supplies, coupled with the skill of Walden and Agnew in working out a temporizing settlement in Batavia thus prevented a serious confrontation with Japan over Indies oil in the fall of 1940.

CIRCUMSTANTIAL evidence is strong that another, hidden, factor had also been at work. Almost unnoticed amid the high level pyrotechnics, a shipping crisis had developed in the Pacific in the fall of 1940. Since the outbreak of war in Europe, German and Italian merchantmen had been barred from the Pacific, and British vessels (almost one-third of the world's total) had been under strict strategic control. Hitler's invasion of Scandinavia and the Low Countries in 1940 swept much of the previously neutral shipping from the seas or into the British system of control, and by late 1940 Britain was concentrating its own merchant marine heavily in the Atlantic. The scarcity thus created in the Pacific hurt Japan badly, and it appears to have been the *real* obstacle to strategic stockpiling.[75] Despite the run on American gasoline, *total* Japanese procurement of petroleum and products from the United States actually declined from 33.2 million barrels in 1938, to 29.9 million in 1939, to 24.9 million in 1940, to 12.6 million in the first seven months of 1941.[76] Of more immediate signifi-

[75]Daniel Marx, Jr., "Shipping Crisis in the Pacific," *Far Eastern Survey*, May 5, 1941, pp. 87–94. By November 1940, this limitation on Japan's ability to stockpile was well known among oil specialists; letter, Walden to Sadler, November 4, 1940, copy in RG 59, 894.24/ 1159; letter, Foote to SecState, November 13, 1940, RG 59, 756D. 6363/103; telegram, Foreign Office to Butler (in Washington), November 15, 1940, FO371(1940)W11699/9160/49; letter, Starling (Petroleum Department) to Steel (Foreign Office), December 19, 1940, FO371(1940)W11898/9160/49.

[76]See Appendix B, Table B-4. Largely because of the drain of the war in China, total Japanese stockpiles also declined from 51.3 million barrels in April of 1939, to 49.6 million in April of 1940, to 48.9 million in April of 1941 (Appendix B, Table B-7).

cance for the turn of events in Batavia, however, was Japan's
subsequent inability to find sufficient tankers to handle the full
quantities under contract. In the nine months before all shipments were halted in August 1941, Japan was only able to lift
half of the promised quantities.

TABLE V — 2

STANVAC/SHELL DELIVERIES AGAINST BATAVIA CONTRACTS
NOVEMBER 1, 1940 THROUGH AUGUST 5, 1941
(In long tons of 2,240 lbs.)

	Contract Quantities	Quantities Delivered	Balance Undelivered
Crude oil (12-month contract)	760,000	441,197	318,803
Products (6-month contract, renewal)	595,500	237,430	358,070
Total[a]	1,355,500	678,627	676,873

[a]Does not include quota sales, for which total deliveries were unavailable.
SOURCE: Contract quantities, Table V-1; deliveries, letter, Walden to W. D.
Crampton (Office of Petroleum Coordinator), September 29, 1941, Box 474,
File 723.3, RG 31.

In retrospect, it would hardly have been worthwhile for the
Japanese to have pressed for quantities they could not handle,
and as long as the threatened embargo did not materialize in
the United States, Japan's shipping capacity was taxed to the
extreme.

LITTLE had occurred, however, to ease Walden's concern for
the safety of his Indies operation. Japanese saber rattling continued unabated, it became increasingly apparent that Anglo-
American governmental interest was concentrated in the
Atlantic rather than the Pacific, and the tight consensus and
administrative control that had enabled Stanvac, Shell, the
Foreign Office, and the State Department to avoid a confron-

tation in 1940 began to erode as the year ended. With victory assured in the Battle of Britain, the Foreign Office began to return to its aggressive stance, and advocates of an embargo in the American government began to plant seeds that would eventually destroy the administrative control so carefully defended by Hull in 1940. As events moved into 1941, Stanvac found itself increasingly on the periphery of American decisions, and the Batavia settlement proved to be the last instance in which the company played a central role in the formulation or execution of American policy in Asia. Walden's intense involvement in oil questions continued unabated, but as more and more agencies entered the act, the decision-making process became increasingly complex and diffused.

SIX

UNTIL PATIENCE RUNS OUT

ALTHOUGH it appeared to be a sudden reversal in policy, the *de facto* embargo on oil created by the freezing of Japanese funds in the United States in July 1941 actually came as the result of a long period of gestation in British and American bureaucracies dedicated to restraining Japan but riddled with disagreement over the most effective means. It is true that the Japanese occupation of southern Indochina triggered economic sanctions, but pressure for a partial or complete oil embargo had been building all spring from a variety of sources. And contrary to the impression created by the manner in which news of the financial freeze was broken to the press, it was *not* Roosevelt's original intent to use the financial freeze to cut off Japan's supply of American oil. The *de facto* embargo was actually the product of a month-long bureaucratic tangle which was not converted into firm policy at the upper echelons of the American government until mid-September. Stanvac was on the periphery of events leading to the July decision and deeply enmeshed in the August tangle, but by mid-1941 it was no longer in center stage on Asian oil. To view its role in proper perspective it is now necessary to step back and examine a variety of forces operating in 1941 that let toward a termination of Japan's oil supply.[1]

WITHIN THE Department of State, Hornbeck had long favored a firm policy toward Japan, but it was constant British prod-

[1] Since this study focuses on Stanvac, only those aspects of Anglo-American policy formulation affecting Asian oil have been reconstructed, leaving numerous critical events untouched—such as the Walsh-Drought mission, the Hull-Nomura talks, and the proposed Konoe-Roosevelt meeting. In the author's opinion, Feis' *Road to Pearl Harbor* remains a highly reliable guide to the broad trend of these events as seen from the American perspective, even though this study differs with Feis on some points and expands on others.

158

ding and overzealous Japanese purchasing that by mid-April finally caused many of his colleagues to join the movement for tighter export controls. Walden was fully aware of this trend, and it appears that the only block to some type of State Department action in April 1941 was fear that it might disrupt the second round of Stanvac and Shell negotiations with the Japanese in Batavia.

British pressure began as an effort to curb Japanese stockpiling by quietly controlling the world's non-Japanese tanker fleet, an idea born in the British War Cabinet's Interdepartmental Committee on Far Eastern Affairs in November 1940.[2] Having weathered Hitler's *Luftwaffe* and averted a threatened invasion, the British had begun to recover their usual toughmindedness in the Pacific. Not content to let the general shortage of shipping take care of the problem, they carefully calculated that if *all* non-Japanese tankers were withdrawn, indigenous production and the volume of imports that could be handled by Japanese merchant and naval tankers would be slightly less than annual consumption, thus precluding any increase in Japanese reserves.[3] While the figures might be debatable, the general idea was clear, and the Embassy in Washington was asked to explore the idea with the Department of State. The United States Maritime Commission had withdrawn American-flag tankers from the Japanese trade in July 1940, except for those owned by American companies carrying their own oil to their own affiliates, but the Commission had no

---

[2]This committee, headed by Parliamentary Undersecretary of State for Foreign Affairs R. A. Butler, was charged with developing recommendations on Far Eastern diplomatic, economic, and military policy designed to "facilitate resistance to Japan and diminish her war potential" while avoiding risk of war for as long as possible; Medlicott, II, 68.

[3]Telegram, Foreign Office to Nevile Butler (British Chargé in Washington), November 15, 1940, with attached Foreign Office minutes, FO371(1940)W11699/9160/49; letter, British Coordination Center (Washington) to Ministry of Economic Warfare (London), December 4, 1940, FO371 (1941)W319/54/49; letter, Starling (Petroleum Department) to Steel (Foreign Office), December 6, 1940, FO371(1940) W11898/9160/49.

legal authority for curtailing the use of American-owned tankers sailing under foreign flags.[4] All the British sought was agreement that both governments would simultaneously exert pressure on the owners of the few foreign-flag tankers still sailing for Japan to have them withdrawn from the trade. The proposal went to Hornbeck on November 20 and disappeared for several months of careful "consideration."[5]

The general issue was resurrected in January when the Japanese government offered Stanvac and Shell substantial increases in their quota business, for which the companies had always provided tankers. As soon as word of this proposal reached the British Embassy, Nevile Butler, the British Chargé, talked with Hornbeck, urging that the American government advise Stanvac to reject the offer since acceptance would be inconsistent with the tanker policy recommended the previous November. Although he gave Butler a noncommittal reply, Hornbeck immediately consulted Welles and Hamilton and then telephoned Walden and Parker on January 8 to work out a quiet arrangement. Since Stanvac and Shell had pressed for years for quota increases, it was agreed that they would formally accept, but to keep in line with Anglo-American policy they would subsequently find it impossible to locate additional tankers. For the time being, then, Stanvac and Shell continued their

[4]Telegram, Lord Halifax (Washington) to Foreign Office, January 26, 1941, FO371(1941)W412/54/49. According to this report, unobtrusive action had been taken under a legal requirement for Maritime Commission approval whenever American-flag vessels were chartered to foreigners. The Commission construed tankers sailing for foreign ports as operating under "charter" unless a corporation was shipping its own oil in its own ship to its own affiliate, and had granted no approval for "chartered" voyages to Japan since July 1940. Confirmation of this action is contained in Hornbeck's conversation with Butler and Hill on January 26, 1941, RG 59, 894.24/1306.

[5]Memorandum of conversation, Hornbeck with Butler and Hill (British Embassy), January 6, 1941, RG 59, 894.24/1306; letter, Butler to Hornbeck, January 8, 1941, FRUS(1941), IV, 775–76; and telegram, Butler to Foreign Office, January 9, 1941, FO371(1941) W412/54/49. Why the normally belligerent Hornbeck chose to bury this recommendation is not clear from the available record.

actual trade just as it had been, with Japan stretching its tanker fleet to lift its f.o.b. purchases from the Indies and the companies providing a limited amount of tankerage to maintain their c.i.f. quota business at its normal volume. Stanvac itself had only one Swedish-flag and one Panamanian-flag tanker on the Indies-Japanese run at the time.[6]

There was one unexpected byproduct to this incident. In their conversation with Hornbeck on January 8, Walden and Parker had pointed out that much of the American petroleum going to Japan was actually being shipped aboard normal freighters in drums, and if the government really "wished to prevent accumulation by the Japanese of huge reserve supplies of oil, this question of drums was more important than the question of the usage of tankers."[7] When heavy usage of drums was confirmed from several other sources, Hornbeck obtained the approval of the Secretary of State for a request to the Administrator of Export Control that metal drums for petroleum products be added to the control list. Maxwell's office readily agreed, and to avoid excessive publicity buried "drums" in a long list of metal products placed under control for defense purposes on February 4.[8] The British still had no answer on tankers, but with Walden's indirect assistance they had helped generate another obstacle to Japanese stockpiling.

[6]Memorandum of conversation, Hornbeck with Butler and Hill, January 6, 1941, RG 59, 894.24/1306; memorandum of telephone conversation, Hornbeck with Walden and Parker, January 8, 1941, RG 59, 894.6363/370; and telegram, Butler to Foreign Office, January 9, 1941, FO371(1941)W412/54/49. The fact that Stanvac continued its quota business at its normal level is confirmed by the figures in a letter, Walden to W. D. Crampton (Office of Petroleum Coordinator), September 29, 1941, Item 4, Box 474, File 723.3, RG 31. The incident is mentioned in Feis (pp. 158–59), but not in its full context.

[7]Memorandum dated January 8, 1941, RG 59, 811.20(D)Regulations/1153.

[8]Memoranda, Hornbeck to Hull, January 14 and 18 (two), 1941, Box 67, "Petroleum" file, Hornbeck MSS; memorandum, Hornbeck to Hull, January 23, 1941, FRUS(1941), IV, 783; letter, Green to Maxwell [January 23, 1941], FRUS(1941), IV, 783–84; Executive Order No. 8669, February 4, 1941, FRUS(Japan, 1931–41), II, 243–48.

The issue continued to simmer in London. During February, the Foreign Office, the Ministry of Economic Warfare, and the War Cabinet's Far Eastern Committee went through an intensive review of measures that could be adopted to curtail Japanese stockpiling,[9] and low-key but persistent approaches began to be made to enlist American cooperation. Rear Admiral Robert L. Ghormley, the American special naval observer in London, forwarded a copy of a British position paper on the "Japanese Oil Situation" to the Chief of Naval Operations on February 11, recommending careful consideration of a suggestion that Japanese stockpiling be curbed by Anglo-American tanker control.[10] A copy of the same position paper was given Hornbeck by the British Embassy on February 14,[11] and on March 3 Lord Halifax, who had replaced Lothian as British Ambassador, presented a formal copy to the Secretary of State.[12] Like Hornbeck, Hull was noncommittal, but sentiment for adopting the proposal had finally taken hold within the Department, and without informing Halifax, Hull initiated a request to the Maritime Commission that pressure be brought on American companies operating foreign-flag tankers to with-

[9]Minutes of meeting at the Ministry of Economic Warfare, February 7, ·1941; minutes of War Cabinet Far Eastern Committee, February 13, 1941; and Foreign Office minutes of February 14 through 26, 1941; all in FO371(1941)W1427/54/49.

[10]Letter, Ghormley to CNO, February 11, 1941, file (SC)JJ7-3/EF37, CNO Records. This letter was used by Herzog in his "Influence of the United States Navy in the Embargo of Oil to Japan" (p. 324) to infer internal Naval opposition to Stark's cautious embargo position. The proposal that Ghormley forwarded, however, was far less than a full embargo, embodying ideas adopted six weeks later by the ultracautious Department of State. No documents could be located indicating what action, if any, the Chief of Naval Operations took on Ghormley's recommendation.

[11]Memorandum of February 14, 1941, Box 67, "Petroleum" file; Hornbeck MSS.

[12]For Hull's memorandum of his conversation with Halifax on March 3, 1941, see *FRUS(1941)*, IV, 788–91; the detailed position paper was not printed, but was located in RG 59, 894.24/1262.

draw them from the Japanese trade.[13] The British were unable to obtain official confirmation of this policy, but noted the drop-off in ship sailings and took parallel action. By late spring there were almost no non-Japanese tankers left in the trade.[14]

This might have ended the debate except that the ferment had by this time created a more complex issue. The essential British argument was for blocking accumulation of additional reserves, and this proposal invariably led to detailed estimates of present reserves, rates of importation of various products, and estimates of civilian and military consumption. British calculations, included in the data Halifax gave Hull on March 3, placed Japanese reserves at 39.5 million gallons, which the British thought would last nine months in a full-scale naval war.[15] Conflicting figures reaching the State Department in March from the Office of Naval Intelligence estimated Japanese reserves at 75.5 million gallons and forecast that this would be adequate for eighteen months of full-scale war.[16] But both

[13]Letter, Breckinridge Long (Assistant Secretary of State) to Admiral Land (Maritime Commission), March 26, 1941, *FRUS (1941),* IV, 800. The ultra-cautious Maxwell Hamilton had recommended adoption of this policy as early as January 23, after determining that only seven American companies would have to be contacted to reach the owners of all 217 American-owned foreign flag tankers. These were: Atlantic Refining, Gulf Oil, Socony-Vacuum, Standard of California, Standard of New Jersey, Texas, and Union Oil. The other 388 American-flag tankers were already under Maritime Commission legal exclusion from the Japanese trade (memorandum, Hamilton to Welles, January 23, 1941, RG 59, 894.24/1309).

[14]Telegram, Halifax to Foreign Office, April 11, 1941, FO371(1941) W3642/54/49; and telegram, Foreign Office to Halifax, April 26, 1941, FO371(1941)W4956/54/49.

[15]Memorandum on the "Japanese Oil Situation," dated February 9, 1941, RG 59, 894.24/1262. This figure was actually calculated for the end of 1939, but the memorandum treated it as a good approximation for the spring of 1941.

[16]Feis, p. 159, n. 14. The original of this ONI report could not be located in the State Department decimal file (RG 59), but *internal* evidence strongly suggests that the report mentioned by Feis was the same as an ONI report dated March 1941, in folder "53.a. (38) [a-d],

estimates calculated Japanese reserves as on the increase—
especially in grades of aviation gasoline just below 87 octane.
Neither assessment was correct; postwar research revealed that
Japan's actual reserves as of April 1, 1941, came to 48.9
million gallons, which the Japanese considered sufficient for
eighteen months under wartime conditions, and the shipping
shortage had already begun to create a slow *decline* rather than
an increase in total reserves.[17] Decisions, however, are made on
perceptions of reality rather than reality, and the ONI and
British estimates increased pressure within the usually cautious
Department of State to curtail what appeared to be excess
accumulation of stocks. This may have been one reason Hull
capitulated on tanker control, but the issue was more complex. Halifax was now suggesting Anglo-American control
over the types as well as the quantity of petroleum products,
with the use of export licensing, preemptive buying, and curtailment of Japan's foreign exchange as well as tanker control.[18]

The risk of a misstep was high, since acceptance of the
ONI data would have required reductions in exports of 10
million more barrels per year than would have been required
by the British estimates[19]—and an error could produce an unplanned confrontation over the Indies. A further worrisome
mirage was created by requests for export licenses, which continued to mount despite the fact that actual exports had slowed.

Oil," Record Group 243, Records of the United States Strategic
Bombing Survey, National Archives, Washington, D.C., hereafter
cited as "RG243." An earlier version of the same ONI Report, dated
August 24, 1940, is located in Vol. 295, pp. 228–45, Morgenthau Diary.

[17]See Table B-7, Appendix B. For the Japanese estimate of how
long their reserves would last in the event of full-scale war, see the
report to the Emperor by Admiral Nagano Osami on July 31, 1941,
as recorded in the *Kido Diary*, p. 296; and translation of documents
attached to the minutes of the Japanese Imperial Conference of September 6, 1941, Ike, p. 154.

[18]Memorandum of conversation, Hull with Halifax, March 3, 1941,
FRUS(1941), IV, 788–91; telegram, Halifax to Foreign Office, March
4, 1941, FO371(1941)W3114/54/49.

[19]Memorandum, Hornbeck to Acheson, March 21, 1941, "Petroleum" file, Box 67, Hornbeck MSS.

In the sensitive area of gasoline below 87 octane, approximately 4 million licensed gallons had left the United States for Japan between July 1940 and March 1941, licenses had been issued for another 5 million gallons not yet exported, and applications were pending for an additional 2 million gallons. Taken together these figures totaled more than twice Japan's annual prewar gasoline imports from all sources, and even the cautious Far Eastern Division began to consider the situation absurd. The department, therefore, began work on some type of quota system based on a defensible criteria such as the level of shipments in the "normal" year of 1936.[20] To provide time for such a system to be perfected, the Department and the Administrator of Export Control on March 6 suspended all further action on pending applications—without public announcement. Since shipments continued for the time being under previously validated licenses, the action caused little comment, but as matters turned out (except for two minor special cases) the Department actually issued no new licenses for gasoline, lubricating oil, or crude to Japan in the five months preceding the *de facto* embargo.[21] By early April the Division of Controls had obtained agreement in principle from Hamilton and Hornbeck to a quota system based on pre-1937 volumes and a stricter definition of aviation gasoline, but the entire project was tabled at this point to avoid complicating Stanvac and Shell's renegotiation of their six-month Batavia contracts, due to expire April 30.[22] British prodding and Japanese zeal had generated a mood

[20]Memorandum by Cabot Coville (Far Eastern Division), March 8, 1941, RG 59, 894.6363/376-1/2; memorandum of April 4, 1941, *FRUS(1941)*, IV, 803–805; memorandum by Yost (Division of Controls), April 9, 1941, *FRUS(1941)*, IV, 805–808.

[21]Letter, Leonard H. Price (Division of Controls) to Maxwell, June 21, 1941, *FRUS(1941)*, IV, 822; letter, Walden to Hornbeck, July 10, 1941, RG 59, 894.24/1559; letter, Green to Acheson, July 19, 1941, RG 59, 811.20(D) Regulations/3884-1/2.

[22]Telegram, Halifax to Foreign Office, March 16, 1941, and telegram, Foreign Office to Halifax, March 23, 1941, both in FO371 (1941)W3208/54/49; memorandum of conversation, Yost with J. S. Dent (Second Secretary of the British Embassy), March 25, 1941,

for tighter controls in the normally cautious Department of State, but the specter of a Japanese move south again blocked action.[23]

ALL SPRING Stanvac and Shell had treated the Batavia contracts with great care, and the stage was set for quiet renewal when the six-month portions expired. Mukai had left for Japan in late November, and his replacement, Ito Yosabura, proved to be more of an irritant to be placated than a practical negotiator.[24] In February, the United States Navy had contacted Stanvac in Manila with a request to purchase oil products, including 92-octane aviation gasoline, from the Indies to save the long tanker haul from California. Since the company had told the Japanese just three months earlier that its entire production of aviation gasoline was already contracted for, Walden promptly checked with the Department of State to determine whether political or military considerations should prevail. Hamilton and Hornbeck discussed this delicate question with Captain R. E. Schuirmann in the Office of the Chief of Naval Operations, but offered no formal objection, and Stanvac agreed

---

FRUS(1941), IV, 799–800; letter, G. F. Thorold (British Embassy) to Hornbeck, March 26, 1941, "Petroleum" file, Box 67, Hornbeck MSS; memorandum, Green to Acheson, April 4, 1941, memorandum, Hawkins to Acheson, April 4, 1941, unsigned memorandum on "Control of Gasoline Exports to Japan," April 4, 1941, all in "Petroleum" file, Box 67, Hornbeck MSS; memorandum by Yost on "The Export of Petroleum Products to Japan," April 9, 1941, *FRUS(1941)*, IV, 805–808; telegram, Halifax to Foreign Office, May 18, 1941, FO371 (1941)W4567/54/49; memorandum, Sumner Gerard (Export Control) to Maxwell, June 17, 1941, Item 97, Box 225, RG169.

[23]Why this moratorium lasted five months is not made explicit in the surviving records, but the existence of adequate previously validated licenses removed the urgency for action, and circumstantial evidence suggests that Hull wanted to avoid any hint of provocation during his private talks with Ambassador Nomura, which began March 8 (Feis, pp. 171-79, and especially 199).

[24]Telegram, H. A. van Karnebeek (Stanvac, Batavia) to Walden, November 23, 1941, and letter, van Karnebeek to Walden, January 3, 1941, both in RG 59, 856D.6363/809.

to sell to the Navy with the proviso that the sale be given minimum publicity.[25] Walden had also kept Hornbeck up-to-date on the status of Stanvac and Shell deliveries from the Indies, so the Department was well aware in April that Japan had been able to find only enough shipping space to lift about two-thirds of the refined products covered in its first six-month contract. It therefore came as no surprise to anyone when the Japanese government on March 30 approached Stanvac and Shell in Yokohama for simple renewal of the contracts for another six months, plus carryover of the full undelivered quantities from the original contracts.[26] With American supplies still available, there was no need to press for more oil than could be transported.

This time a consensus existed among Stanvac, Shell, and the American, British, and Dutch diplomatic establishments that quantities should be kept to the minimum possible without provoking a crisis. The British deferred to the Department of State for an initiative, and Hornbeck suggested to Walden that the companies agree only to a six-month renewal of the contracts under exactly the same terms as before, temporarily leaving open the question of how much of the first contract would be carried over. Stanvac and Shell agreed and were surprised to find that the Japanese offered minimum protest, although formal signing of the new contracts in Yokohama was delayed until May 15. Subsequent negotiations set the carry-

[25]Memorandum of conversation, Alger Hiss (Assistant to Hornbeck) with Walden, February 17, 1941, RG 59, 856D.6363/810; letter, Parker to Hornbeck, February 17, 1941, and memorandum by Hamilton, March 6, 1941, both in RG 59, 856D.6363/811.

[26]Letter, Bridgeman (Petroleum Department) to Steel (Foreign Office), April 1, 1941, FO371(1941)W3890/54/49; letter, Hornbeck to Walden, April 3, 1941, and Hiss memorandum of April 4, 1941, both in "Petroleum" file, Box 67, Hornbeck MSS; Walden's referenced letter of April 2 to Hornbeck could not be located, but a later version of the delivery statistics is contained in a letter from Walden to W. C. Crampton (Office of the Petroleum Coordinator), September 28, 1941, Item 4, Box 474, File 723.3, RG 31; see also telegram, Foreign Office to Halifax, April 5, 1941, FO371(1941)W3208/54/49.

over at only a fraction of what had not been delivered by April 30.[27] The combined volume of gasoline, kerosene, diesel, and fuel oil under the original Stanvac and Shell Batavia contracts had been 323,250 tons, of which only 181,795 tons had been shipped, leaving 141,455 tons undelivered. The renewal contracts again called for 323,250 tons in six months, with only 36,000 tons added from the previous contract.[28] All parties, however, appeared satisfied with this resolution of an issue that had generated extreme tension only six months earlier. An uneasy equilibrium had been established by the shortage of tankers and American restraint.

WITHIN THE American government, however, this restraint came almost exclusively from the Department of State, which at this point still retained effective administrative control over oil policy. Even though his lieutenants had become restive, Hull was determined to avoid provocations that might result in a Japanese descent on the Indies, and in this course he had the firm support of the President. Behind the floodwall, however, pressure for more vigorous action had been building in several other agencies. Ever since he reentered the Cabinet as Secretary of War in July 1940, Stimson had been carrying on a private campaign for economic warfare in general and sanc-

[27]Telegram, Foreign Office to Halifax, April 5, 1941, FO371(1941) W3208/54/49; letter, Walden to Hornbeck, April 10, 1941, "Petroleum" file, Box 67, Hornbeck MSS; telegram, Halifax to Foreign Office, April 11, 1941, FO371(1941)W4567/54/49; two memoranda of conversations, Hornbeck with Walden, April 15, 1941, both in "Petroleum" file, Box 67, Hornbeck MSS; letter, Walden to Hornbeck, April 28, 1941, RG 59, 856D.6363/815; letter, Bridgeman (Petroleum Department) to Ashley-Clarke (Foreign Office), April 29, 1941, FO371(1941)W5657/54/49; letter, Parker to Hornbeck, May 15, 1941, and memorandum of conversation, Hiss with Walden and Parker, May 19, 1941, both in RG 59, 856D.6363/814. The settlement is accurately summarized in Medlicott, II, 84, but apparently was overlooked by Feis, who only noted the continuing Japanese rhetoric in the regular Batavia talks (Feis, p. 189).

[28]Letter, Walden to W. D. Crampton (Office of the Petroleum Coordinator), September 29, 1941, Item 4, File 723.3, Box 474, RG 31.

tions against Japan in particular.[29] Convinced that historically the Japanese had demonstrated they would yield when faced with "clear language and bold actions" by the United States, Stimson believed that "soft words" would only "encourage Japan to bolder action."[30] For this reason he consistently argued for a full petroleum embargo at every opportunity. Of more practical importance, however, was the nucleus of an economic warfare staff which Stimson established as a "study group" in the Army Industrial College and in December converted into an obscure "Projects Section" in Maxwell's Export Control Administration.[31] By early May this group, working quietly through interdepartmental committees, had produced and circulated no less than eighteen contingency plans on *how* to cripple Japan economically, with only passing reference to *whether* such a policy should be adopted.[32] In effect, Stimson

[29]Henry L. Stimson and McGeorge Bundy, *On Active Service in Peace and War* (New York: Harper, 1948), pp. 382–87; Elting E. Morison, *Turmoil and Tradition: A Study in the Life and Times of Henry L. Stimson* (Boston: Houghton Mifflin, 1960), pp. 522–23; Richard N. Current, *Secretary Stimson: A Study in Statecraft* (New Brunswick: Rutgers University Press, 1954), pp. 144–47.

[30]Stimson memorandum of October 2, 1940, following diary entry of the same date, Stimson MSS.

[31]Memorandum of February 10, 1941, "Embargo" folder, Safe File, Record Group 107, Records of the Office of the Secretary of War, National Archives, Washington, D.C., hereafter cited as "RG 107"; Johns, "Historical Record of Export Control," pp. 119–35, Item 145, Box 887, RG 169. Because of State Department objections to official liaison between this group and the British, members of the Project Section met secretly at night with representatives of the British Embassy in Washington, and Maxwell had one of his own staff assigned as military attaché in London for private liaison with the Ministry of Economic Warfare.

[32]These included one special study on the "Japanese Oil Situation," the first draft of "A Co-ordinated Plan of Economic Action in Relation to Japan," and sixteen studies on Japanese economic vulnerability in different commodities, one of which was petroleum; letter, Thomas Hewes (Chief, Projects Section) to Maxwell, March 31, 1941, "Projects Section Correspondence" folder, "Administrator of Export Controls, Central File," Item 97, Box 78, RG 169; Report number CC6-1,

had institutionalized his views, and difficult as it would be to assess the actual impact, it is hard to believe that this move did not contribute to a more aggressive mood at the working level of those agencies in close contact with the question.[33]

Meanwhile, Morgenthau had been at work on another tack. In late December he had proposed a sweeping Executive Order freezing *all* foreign funds in the United States and creating an Economic Defense Board headed by the Secretary of State to guide the administration of this program along with export control. Actual execution of the freezing order would be handled by the Treasury. With experience in such administrative devices, Hull viewed this suggestion as a thinly disguised attempt to capture control of foreign economic policy and refused to concur in the plan as it stood. As a result, the idea for an Economic Defense Board went to the Budget Bureau for a lengthy attempt to reconcile differing viewpoints, and by June had almost reached maturity, with hawkish Vice President Henry Wallace proposed for the chairmanship.[34] The sug-

---

"The Economic Vulnerability of Japan in Petroleum," by the Interdepartmental Committee on Petroleum, April 1940 [sic], Item 158, Box 963, RG 169; letter, Hewes to Maxwell, May 1, 1941, Item 171, Tab G4, Box 995, RG 169; letter, Maxwell to Executive Secretary, Army and Navy Munitions Board, May 9, 1941, "Projects Section Correspondence" folder, "Administrator of Export Control, Central File," Item 97, Box 78, RG 169.

[33]In June, Maxwell's office revised and codified its export regulations on petroleum, but State still controlled oil policy, and the revised regulations did not significantly change the definition of aviation gasoline prohibited to Japan (*Documents on American Foreign Relations*. III, 485–86; Export Schedule No. 10, effective June 20, 1940, Item 171, Tab B9, Box 169, RG 169; Directive No. 427 on "Petroleum Products and Tetraethyl Lead," June 27, 1941, Item 177, "Export Licensing Memoranda, 1940–42," Box 1021, RG 169).

[34]Feis, pp. 142–44; memorandum, Hornbeck to Hull, January 7, 1941, unmarked box: "Memoranda, Dispatches and Letters, 1941," Envelope 48, Hornbeck MSS; letter, Morgenthau to Wallace, July 15, 1941, with enclosures describing the evolution of the idea, Vol. 421, pp. 144–72, Morgenthau Diary; Johns, "Historical Record of Export

gestion for financial controls fared somewhat better, with administrative responsibility assigned to the Treasury and a freeze imposed on most Axis funds on June 14. Policy guidance was to come from a three-man State-Treasury-Justice Committee, with State represented by newly appointed Assistant Secretary Dean Acheson, an ideological ally of Stimson and Morgenthau on Asian matters.[35] Japan was carefully excluded from the freezing order, but the machinery was in place and additional countries could be added with the stroke of a pen. If there is such a phenomenon as bureaucratic momentum, it was clearly in the direction of economic sanctions, with Hull and Roosevelt's reluctance to provoke a Japanese attack on the Indies as virtually the only check.

ALL THESE pressures, of course, were in response to a steadily worsening world situation, and by late spring Stanvac, Shell, State, and the Foreign Office found themselves and their policy of relative restraint increasingly shunted to the sidelines by events outside their control. As the war in Europe continued, the initiation of joint military planning and the passage of Lend Lease in the spring of 1941 demonstrated an increasing American conviction that its own security was inextricably linked with that of Britain. A strategic decision had also been made to place first priority on Europe and avoid—or at least delay

---

Control," pp. 136–46, Item 145, Box 887, RG 169. By early May Wallace had become convinced that petroleum shipped to Japan might ultimately be used against the United States, and believed that the time was "ripening fast . . . for more complete action in shutting off exports to Japan" (letter, Wallace to Hull, *FRUS(1941)*, IV, 815–16).

[35]Executive Order No. 8785, June 14, 1941, greatly expanded the list of European countries whose American funds had been originally frozen as a safeguard when they were invaded by Germany; Denmark and Norway were the first; their funds had been frozen by Executive Order No. 8389, April 10, 1940; *Documents on American Foreign Relations*, II, 540–51, and III, 537–41; on composition of the committee, see Dean Acheson, *Present at the Creation: My Years in the State Department* (New York: Norton, 1969), pp. 21–23.

for as long as possible—a crisis in the Pacific.[36] It was at least partly for this reason that Roosevelt and Hull agreed in March to convert a private initiative by two Catholic missionaries into secret negotiations between Hull and Ambassador Nomura Kichisaburō in an attempt to check an ominous drift toward open conflict. In retrospect, the optimism these talks generated in both Washington and Tokyo appears misplaced, since each side misconstrued the other's position as more flexible than was actually the case. Reduced to essentials, Japan wanted the United States to discontinue its support of China, and the United States wished Japan to return to the conditions that prevailed in 1937.[37] These differing objectives provided no immediate cause for war, but the nagging discussion of China distracted attention from the strategic conflict developing in Southeast Asia and made resolution of other issues all the more difficult.

Another issue arose almost immediately. On June 17 the nine-month Japanese economic discussions with the Netherlands Indies were finally broken off with no significant results except the oil contracts,[38] and on June 22 Hitler altered the

[36]The Lend Lease Act was passed on March 11, and Anglo-American staff talks produced the first joint strategic plan, ABC-1, on March 27; the supporting American war plan, RAINBOW 5, was completed May 14; Feis, pp. 153–55, 165–70; Langer and Gleason, *Undeclared War*, pp. 213–89; Matloff and Snell, pp. 32–48.

[37]Feis, pp. 171–79, 192–95, 199–201, 203–205, 211–12, 222, 248–50, 252, 272–78, 303–305, 312–21; Langer and Gleason, *Undeclared War*, pp. 313–15, 320–21, 464–85, 631–45, 656–62, 675, 723, and 836–941; Robert J. C. Butow, "The Hull-Nomura Conversations: A Fundamental Misconception," *American Historical Review*, LXV, No. 4 (July 1960), 822–36.

[38]Kobayashi had been replaced on December 28 by Yoshizawa Kenkichi, who spent all spring in Batavia attempting to negotiate far-reaching economic concessions. The Dutch refused to be incorporated into the "Greater East Asia Co-Prosperity Sphere," and the Japanese finally broke off the talks after concluding that economic autonomy could not be achieved by diplomatic methods; Van Mook, pp. 77–123; Ike, pp. 47–51; Feis, pp. 207–208; Lu, pp. 152–53; Langer and Gleason, *Undeclared War*, pp. 482–83; Jones, *Japan's New Order*, pp. 259–60.

strategic balance by launching his attack on Russia. It is difficult to gauge the impact of these two developments on Japanese thinking, but they came at a time when Tokyo was already reassessing its "southern policy," and they clearly accelerated that process. At least as early as May 22 the Japanese government had begun to despair of achieving economic autonomy by peaceful means in Southeast Asia and had commenced discussion of a possible "resort to armed measures in the South."[39] Both the Army and Navy favored at least strengthening their position by the establishment of bases in southern Indochina, but after advance rumors of German plans to invade Russia reached Tokyo on June 6, Foreign Minister Matsuoka began to argue vehemently for a Japanese move against Siberia instead. Through the rest of June, the debate ranged over whether to move north or south or in both directions, and when the decision to occupy only southern Indochina was formally ratified at an Imperial Conference on July 2 it appeared the epitome of caution. Incredibly, the possibility of this move provoking Anglo-American economic sanctions received only scant attention, and there was no serious discussion of what to do if that occurred except to "act resolutely."

From the American viewpoint the Imperial Conference decision of July 2 was the last straw. Through the highly classified code-breaking operation known as "Magic," Hull, Roosevelt,

[39]Statement attributed to Matsuoka in the Liaison Conference of May 22, 1941; Ike, p. 38. Secondary works vary widely on when and why discussion began on a further move south; this interpretation is based primarily on excerpts of Liaison Conference minutes translated in Ike, pp. 36–90, but has been compared with Feis, pp. 209–18; Langer and Gleason, *Undeclared War*, pp. 625–54; Butow, *Tojo*, pp. 204–20; Jones, *Japan's New Order*, pp. 259–63; Lu, pp. 185–88; Toshikazu Kase, *Journey to the Missouri* (New Haven: Yale University Press, 1950), pp. 47–49; *Konoye [sic] Memoirs*, English translation in U.S. Congress, Joint Committee on the Investigation of the Pearl Harbor Attack, *Pearl Harbor Attack, Hearings* (Washington, D.C.: Government Printing Office, 1946), hereafter cited as "Pearl Harbor Attack," Exhibit No. 173, Part 20, pp. 3985–4029; entries for June 3 through July 24, 1941, in the *Kido Diary;* and Affidavit of Tojo Hideki, pp. 52–63, IMTFE Exhibit No. 3655.

and other key officials had access through most of 1941 to even the most secret Japanese diplomatic messages, and therefore knew the gist of the July 2 decision as soon as it was transmitted to the Embassies in Washington, Berlin, and Moscow.[40] To this warning was added a decoded July 14 intercept of a military message asserting that "after the occupation of French Indo-China, next on our schedule is . . . the Netherlands Indies."[41] Even before it took place the American government saw the impending occupation of southern Indochina as simply a stepping stone for the long feared attack on the Indies. When 40,000 Japanese troops landed at Camranh Bay and Saigon on July 24 with the grudging consent of the Vichy government, American opponents of economic sanctions found the rationale for their position badly undermined. Why continue to supply the sinews of war if a move against the Indies was now inevitable?

The American reaction to the Japanese occupation of southern Indochina was swift and firm, although lacking in precision in the application of economic sanctions.[42] Hull had left for White Sulphur Springs on June 23 to recuperate from an illness, leaving Welles in tactical command at State for the next six weeks. When advance word of Japanese plans reached Washington, Roosevelt authorized Welles to give Halifax explicit assurances on July 10 that if "Japan now took any

[40]Telegram from Tokyo to Berlin and Washington, July 2, 1941, *Pearl Harbor Attack,* Exhibit No. 1, Part 12, pp. 1–2; Feis, p. 219; Langer and Gleason, *Undeclared War,* pp. 636–37; Hull, pp. 1012–13.

[41]Telegram, Canton to Tokyo, July 14, 1941, *Pearl Harbor Attack,* Exhibit No. 1, Part 12, pp. 2–3.

[42]Since this study focuses on Stanvac, the narrative touches only essential points leading to the *de facto* embargo on oil. For a fuller discussion of events in July, see Feis, pp. 227–49; Langer and Gleason, *Undeclared War,* pp. 645–54; and Medlicott, II, 105–18; although those works interpret the origins of the *de facto* embargo slightly differently. A reexamination of primary sources, including material in Ickes' diary, the Stimson papers, and records in Maxwell's office (which appear not to have been used by Feis or Langer and Gleason), suggests a high degree of consistency in Roosevelt's position and the directives actually issued. The real confusion arose in post-freeze administration.

overt step through force or through the exercise of pressure to conquer or to acquire alien territories in the Far East, the Government of the United States would immediately impose various embargoes, both economic and financial, which measures had been under consideration for some time past and which had been held in abeyance for reasons which were well known to the Ambassador."[43] Precisely how far these "embargos" would go still remained in doubt. When the subject came under discussion at a Cabinet meeting on July 18, there was consensus for some type of action, including a freeze of Japanese funds, but on petroleum Roosevelt himself spoke only of a reduction in shipments, since "to cut off oil altogether at this time would probably precipitate an outbreak of war in the Pacific and endanger British communications with Australia and New Zealand."[44] Welles put his staff to work, and by the 21st State and Treasury had tentatively agreed on draft documents to freeze Japanese funds, lower the octane count on permissible gasoline shipments, and establish a reduced quota for all petroleum products based on the "normal" years 1935 and 1936.[45]

On July 24, the day Japanese warships appeared at Camrahn Bay to begin unloading troops, the Cabinet again discussed sanctions. Roosevelt reiterated his decision that the United States would "continue to ship oil and gasoline,"[46] but directed that the freezing order itself be all-inclusive so that "policy can be changed from day to day without issuing any

[43]Memorandum of conversation, Welles with Halifax, July 10, 1941, *FRUS(1941)*, IV, 300–303.

[44]Memorandum, Robert P. Patterson (Undersecretary of War who attended in Stimson's absence) to Stimson, July 18, 1941, Correspondence Box 387, Stimson MSS. Morgenthau's version of this discussion is in essential agreement with Patterson's account: Morgenthau "Presidential Diary," memorandum of July 18, 1941, pp. 946–48. For a fuller account of this Cabinet discussion, see Blum, II, 377–78; and Ickes, III, 583–84.

[45]Hornbeck memorandum of July 19, 1941, RG 59, 894.24/1575-1/2; memorandum, Green to Acheson, July 19, 1941, RG 59, 811.20 (D) Regulations/3884-1/2; memorandum, Foley to Morgenthau, July 21, 1941, Vol. 423, pp. 194–97, Morgenthau Diary.

[46]Ickes, III, 588.

further orders."[47] Administration was to be guided by the Foreign Funds Control Interdepartmental Committee, with licenses granted "for the shipment of petroleum as the applications are presented to the Treasury," although this policy "might change any day and from there on we would refuse any and all licenses."[48] To preserve this internal flexibility there was to be no immediate public announcement on oil policy, and news of the financial freeze was released to the press on the 25th without qualifying clauses—leading to widespread public belief (shared by the Japanese government) that oil shipments had been completely blocked along with other trade.[49] Within the upper echelons of the American government Roosevelt's decision appears to have been clearly understood. On July 25, the Chief of Naval Operations, Admiral Stark, alerted his fleet commanders and their Army counterparts to the impending financial freeze, but noted that "export licenses will be granted for certain grades of petroleum products . . . and possibly some other materials."[50] The public remained under the impression that *all* trade with Japan had been ended.

[47]Memorandum, Daniel W. Bell (who attended in Morgenthau's absence) to Morgenthau, July 24, 1941, Vol. 424, pp. 145–47, Morgenthau Diary.

[48]Ibid. The memorandum does not make clear *who* would decide on such a change in policy, but the context suggests that Roosevelt intended to reserve that decision for himself.

[49]White House press release, July 25, 1941, *FRUS(Japan 1931–41)*, II, 266–67; Executive Order No. 8832, July 26, 1941, *FRUS(Japan 1931–41)*, II, 267; Feis, pp. 242–43; Langer and Gleason, *Undeclared War*, p. 652. Headlines in the *New York Times* of July 26 reported that, "ROOSEVELT FREEZES ALL JAPANESE ASSETS HERE; HALTS OIL SHIPMENTS AND SILK IMPORTS." On July 31, Admiral Nagano Osami, Chief of the Naval General Staff, reported to the Emperor that if "restoration of diplomatic relations [sic] between the U.S.A. and Japan were impossible, we would be cut off from supplies of oil, our store of which would run out in two years . . . [or in case of war] . . . one and a half years." (entry for July 31, 1941, *Kido Diary*).

[50]Testimony of Admiral Stark, *Pearl Harbor Attack*, Part 5, p. 2115. Stark had argued against a complete embargo as late as July 21 and understood that Roosevelt had concurred with this recommenda-

THE FINANCIAL freeze was instituted promptly, but establishing administrative arrangements for the new policy of reduced oil shipments took longer. Acheson's Interdepartmental Committee hammered out the final details, Welles obtained Roosevelt's approval on July 31, and by August 9 Maxwell's Export Control organization had established a highly restrictive new definition for aviation gasoline and detailed export quotas— which, unknown to the general public, remained in full legal effect until after the attack on Pearl Harbor.[51] Outstanding

tion and a similar one made by the Joint Board on July 25; letter, Stark to Hull, July 22, 1941, with enclosed War Plans Division study, July 19, 1941, and letter, Stark to Welles, July 22, 1941, all in *FRUS (1941)*, IV, 835–41; for Stark's additional comment on the Joint Board recommendation, see Charles F. Romanus and Riley Sunderland, *Stilwell's Mission to China*, in the *United States Army in World War II: China-Burma-India, Theatre* series (Washington, D.C.: Department of the Army, 1953) p. 24, note 61.

[51]Memorandum by George F. Luthringer (International Economic Affairs) to Acheson, July 30, 1941, *FRUS(1941)*, IV, 844–46; letter, Welles to Roosevelt, July 31, 1941, with the notation "SW OK FDR," *FRUS(1941)*, IV, 846–48; minutes of meeting of OAEC Subcommittee on Regulations, August 1, 1941, and Export Schedule No. 15 effective August 2, 1941, both in "Subcommittee Meeting" folder, Item 87, Box 699, RG 169; memorandum of meeting between Export Control and Treasury personnel, August 6, 1941, "Japan-Oil Shipments" folder, Box 513, Record Group 131, Alien Property Records, Accession No. 61A109, "Foreign Funds Control—General Correspondence," Federal Records Center, Suitland, Maryland, hereafter cited as "RG 131"; telegram, Halifax to Ministry of Economic Warfare, August 6, 1941, FO371(1941)F7696/1299/23; memorandum from Treasury to State, August 7, 1941, *FRUS(1941)*, IV, 853–55; Directive No. 630, August 8, 1941, Directive No. 632, August 9, 1941, and Licensing Instruction No. 106, August 14, 1941, all in "Export Licensing Memoranda, 1940–42" folder, Item 177, Box 1021, RG 169; memorandum by Yost, August 16, 1941, RG 59, 811.20(D) Regulations/4148-2/3. The Japanese quotas for low-grade crude and refined products for August 1 through December 31 were virtually identical to the quantities recommended by the Division of Controls on July 19 and totaled 10.5 million barrels, slightly less than the 12.6 million barrels actually shipped in the first half of 1941. Together this would have been slightly less than the 24.9 million barrels shipped in 1940, and about equal to the average for 1935–36; see Table B-4, Appendix B.

petroleum export licenses had been revoked August 1 with a cryptic announcement that applications could be resubmitted but would not be approved if they exceeded "prewar quantities" or involved "fuels and oils suitable for use in aircraft and . . . certain raw stocks from which such products are derived."[52] Shippers now had to apply first at the Department of State for an export license (under unannounced criteria established by the semi-independent Administrator of Export Control) and then at the Treasury Department for an exchange permit (under criteria yet to be established by the Interdepartmental Foreign Funds Control Committee), with no advance guidance as to what might be acceptable. To cap this administrative edifice, the long-discussed Economic Defense Board under Vice President Wallace was established with ill defined advisory responsibilities by Executive Order on July 30,[53] and the principal architects—Roosevelt and Welles—left Washington August 3 for secret talks with Winston Churchill at Argentia, Newfoundland. In retrospect, the chief product of Japan's occupation of southern Indochina was not an oil embargo, but an unguided bureaucracy biased at the working level against liberality toward Japan.

ALTHOUGH by mid-September this maze had congealed into a *de facto* embargo, it should be clear from the above account that this end was not Roosevelt's original intent. The initial termination of shipments from both the United States and the Indies was purely the result of a bureaucratic tangle as various

[52]White House press release quoted in telegram, Welles to Grew, August 1, 1941, *FRUS(1941)*, IV, 851. The announcement of revocation and suggestion for resubmission were in a State Department press release also telegraphed to Grew August 1, 1941, *FRUS(1941)*, IV, 851.

[53]Executive Order No. 8839, July 30, 1941, *Documents on American Foreign Relations*, IV, 180–82. There is a useful history of this agency in Franklin D. Roosevelt, *The Public Papers and Addresses of Franklin D. Roosevelt*, compiled and edited by Samuel I. Rosenman (13 vols.; New York: Random House (1–5), Macmillan (6–9), and Harper (10–13), 1938–1950), X, 293–97.

agencies, governments, and companies (including Stanvac) attempted to discern the real intent of American policy and adjust their own actions accordingly.[54] As it turned out, only one small shipment of lubricating oil left the United States for Japan after the revocation of export licenses on August 1, and the last delivery from the Indies was a cargo of crude shipped August 5 under Shell's Batavia contract aboard the Japanese tanker *San Pedro Maru*.[55] All those involved in oil shipments felt that American attitudes were becoming increasingly tough minded, and from August on no one was inclined to unsnarl the tangle and be the first to release a shipment. When it gradually became apparent what had happened, Hull and Roosevelt elected to leave the situation as it was, and in effect ratified the unplanned embargo. Bureaucratic momentum carried the day. Since Stanvac by this time had begun to func-

[54]This interpretaton is at variance with the memoirs of Dean Acheson, who years later wrote that he and Welles decided in late July to tie up Japanese trade by deliberately stalling on the issuance of exchange permits (Acheson, p. 26). Contemporary records of the Foreign Funds Control Committee, however, show that the July discussion between Acheson and Welles was only on technical aspects of coordination with export control, and the decision to use deliberate stalling tactics was not made until September 5, on the basis of discussions with Hull. While Acheson's account could be correct, it appears more likely that he combined the two discussions in his memory, and the interpretation given here is based on the contemporary records (excerpt of memorandum for the Secretary's files of July 30, 1941, and memorandum for the files by N. E. T[owson], September 6, 1941, both in "Japan-Oil Shipments" folder, Box 513, RG 131).

[55]Treasury Department reports on exports of petroleum products for the weeks ending August 2 and 9, 1941, Vol. 427, p. 325, and Vol. 431, p. 232, Morgenthau Diary; letter, Walden to W. D. Crampton (Office of Petroleum Coordinator), September 26, 1941, File 723.3, Item 4, Box 474, RG 31; telegram, Foreign Office to Halifax, September 6, 1941, FO371(1941)F9051/1299/23. The sole shipment from the United States was 1,578 barrels of lubricating oil already loaded aboard the SS *Tatuta Maru* at San Francisco, for which an exchange permit was issued August 2; memorandum by O. A. S[chmidt], August 2, 1941, memorandum by N. E. T[owson], August 4, 1941, and telegram, Slade to Chalmers (Foreign Funds Control), January 6, 1942, all in "Japan-Oil Shipments" folder, Box 513, RG 131.

tion almost as an extension of the American State Department in the implementation of Asian oil policy, it was heavily involved in the expansion of this embargo to the Indies, but its role will be clearer if several parallel developments are mentioned first.

By the spring of 1941 the British government had decided to stay in lockstep with the United States on economic policy toward Japan, and ever since early July it had been struggling to gain a clear idea of American intent so that it could take supportive action. When it finally learned (on the 25th) that a financial freeze was definitely planned, the British government moved with considerable speed to take parallel action on sterling accounts, encouraging the Dominions to follow suit. Thus by the end of July, Japanese trade with all parts of the Empire (except for a few special cases in Borneo and Malaya) had been halted along with that from the United States. From an American viewpoint, this represented admirable cooperation, but the British still had no clue as to how strictly the freeze was to be administered and like everyone else puzzled over cryptic press releases and evasive answers to direct questions.[56]

Part of the puzzlement derived from a strange episode that began August 1 in Washington. Immediately after the revocation of outstanding export licenses, a number of applications had been resubmitted, and on August 11 three of these, which met the new criteria, were approved by the Division of Controls (with Hull's concurrence) for shipment aboard two Japanese tankers then waiting to load at San Pedro, California.[57] The combined value was $178,650, and Counselor Iguchi Sadao

[56]Medlicott, II, 105–17; Feis, pp. 227–50; Langer and Gleason, *Undeclared War*, pp. 645–54.

[57]The three applications included $91,000 to cover a purchase of 90,000 barrels of diesel oil from Standard Oil of California by Asano Bussan; $75,000 to cover a purchase of 75,000 barrels of blended diesel oil from Tidewater Associated by Mitsubishi Shoji Kaisha; and $12,650 to cover a purchase of 700 barrels of turbine oil from Richfield Oil by Mitsui. The shipments were to have been loaded aboard the *Otowasan Maru* and the *Nittiei Maru;* telegram, Hull to Collector of Customs, Los Angeles, August 11, 1941 (initialed by Acheson and personally signed by Hull), RG 59, 811.20(D)Regulations/3912A;

and Financial Attaché Nishiyama Tsutomu of the Japanese Embassy immediately applied at the Treasury for release of this amount from blocked Japanese accounts in American banks to pay the suppliers. They met with an unexpected complication. Ever since Morgenthau first suggested an across-the-board freeze in December 1940, the Treasury had tracked a steady withdrawal of Japanese funds from American banks—presumably in anticipation of just such action—including almost $1 million in currency withdrawn by the Japanese Navy and at least $6 million transferred to dollar accounts in Brazil.[58] Since the Treasury considered the dollar and sterling freeze to be incomplete until these free funds had been used up, Nishiyama was told that the oil should be paid for out of cash withdrawn just before the freeze rather than from blocked accounts.[59]

Iguchi appealed to Acheson with the complaint that the purchases were being made by Mitsui and Mitsubishi while the cash belonged to the Navy, which was reluctant to take directions from civil authorities, and Acheson suggested that they try to transfer funds from unfrozen dollars accounts in Brazil.[60]

memoranda of August 12 and 13, 1941, in "Japan-Oil Shipments" folder; Box 513, RG 131; memorandum, Acheson to Welles, August 16, 1941, *FRUS(1941)*, VI, 858–60; communication from the Japanese Embassy to the Treasury Department, October 9, 1941, *FRUS(1941)*, IV, 895–96.

[58]In the six months prior to the freeze, reported short-term Japanese assets in the United States declined from $124.8 million to $76.3 million; memorandum, Gass to White (Treasury Department), August 11, 1941, copy in RG 59, 840.51 Frozen Credits/3714-2/9.

[59]Memorandum of telephone conversation, J. W. Pehle (Treasury) with Acheson, August 9, 1941, and memorandum of conversation, A. U. Fox (Treasury) and others with Nishiyama, August 15, 1941, both in "Japan-Oil Shipments" folder, Box 513, RG 131.

[60]Memorandum of conversation, Acheson with Iguchi, August 15, 1941, *FRUS(1941)*, IV, 857–58. While traditional Japanese attitudes make Iguchi's explanation plausible, the full story behind this complication would require extensive research in Japanese records, if the relevant ones could be found. It is possible that the initial problem over currency held by the Navy was never known in Tokyo, although later conversations indicate that the question of Brazilian accounts was referred back to the Japanese Foreign Office.

Instead of doing this promptly, the Japanese used up another week in a vain attempt to have funds released from a blocked American account, and then on August 22 agreed to try Brazil.[61] On September 4 Nishiyama reported that he was working on this approach, but it still "might take two or three weeks to arrange the transfer."[62] The Japanese tankers continued to ride at anchor on the West Coast. For more than a month, no oil left the United States for Japan, and there was a deepening impression inside and outside the government that this was the original American intent. On September 5, Hull conferred with Acheson, Hornbeck, and Charles W. Yost of the Division of Controls,[63] and decided to direct Foreign Funds Control to convert the complications that had arisen into deliberate stalling tactics pending further "clarification . . . within the next few weeks."[64] At this point, Stanvac became entangled in the same web, and before continuing the tale of the Japanese licenses it would be best to go back and trace parallel developments involving Stanvac and the Indies.

[61]Memoranda of conversations, Fox with Nishiyama, August 19 and 22, 1941; "Japan-Oil Shipments" folder, Box 513, RG 131; memorandum, J. M. Jones (Far Eastern Division) to Acheson, August 22, 1941, RG 59, 840.51 Frozen Credits/3476.

[62]Memorandum of conversation, Acheson with N. E. Towson (Treasury), September 5, 1941, *FRUS(1941)*, IV, 868–69. See also Towson's memorandum of his September 4 conversation with Nishiyama dated September 6, 1941, "Japan-Oil Shipments" folder, Box 513, RG 131.

[63]Memorandum of telephone conversation, Acheson with Towson, September 5, 1941, *FRUS(1941)*, IV, 868–69.

[64]Towson's memorandum of his September 5 telephone conversation with Acheson, dated September 6, 1941, "Japan-Oil Shipments" folder, Box 513, RG 131. Towson's version of this conversation is far more explicit on the use of stalling tactics than is Acheson's. A close reading of his conversation with Nishiyama that same afternoon suggests that Towson immediately complied; memorandum of conversation, Miller, Fox and Towson with Nishiyama, September 5, 1941, *FRUS(1941)*, IV, 869–70. See also Yost's memoranda to Acheson, September 4 and 5, 1941. "Petroleum" file Box 67, Hornbeck MSS.

IMMEDIATELY after the July 26 American freeze on Japanese funds, the Netherlands Indies had frozen all Japanese guilder accounts and announced that future oil exports would require special permits.[65] Japan likewise froze all American, British, and Dutch yen accounts, thereby requiring a complete renegotiation of terms of trade in an atmosphere that could only be described as tense. All parties, including Stanvac and Shell, looked to the American government for clues as to how the freeze would be administered, but it will be recalled that in late July the United States still had not completed the administrative structure for its unannounced policy to reduce but not terminate oil shipments to Japan. Walden kept in constant contact with the State Department, and immediately after the freeze he called at both Treasury and State to be sure his company's position was fully coordinated and understood. He briefed Foreign Funds Control on the details of Stanvac and Shell financial transactions with Japan and asked for guidance so that "the commercial decisions which his company will be called upon to make will be fully in accord with . . . [American] . . . policy."[66] Stanvac would be perfectly "willing to discontinue sales of oil to Japan" if that were the policy decided upon,[67] or the company would continue to ship, if that were the action desired. But he told Hornbeck that "public criticism of oil shipments to Japan" was now so great that he would need some type of formal approval, even if it was only "an exchange license," rather than continuing to deal on a strictly informal basis. Walden emphasized that Stanvac wanted "to be guided entirely by . . . the policies of the United States Government," and specifically asked "whether his company should apply to the Netherlands East Indies authorities for the recently

[65]Telegram, Grew to SecState, August 26, 1941, *FRUS(1941)*, IV, 281–82; Feis, p. 246.

[66]Memorandum of conversation, Fox and Schmidt (Foreign Funds Control) with Walden and N. T. Singer (Stanvac), July 29, 1941, "Japan-Oil Shipments" folder, Box 513, RG 131.

[67]Memorandum on July 28 meeting with Walden, Ullman to White (Treasury), July 31, 1941, Vol. 426, p. 258, Morgenthau Diary.

required export permits" and whether State would recommend to the Treasury that "the Japanese be allowed to transfer dollars to Standard-Vacuum for such shipments."[68] Even though Japanese tankers were then arriving in the Indies for loading, Hornbeck asked that Stanvac take no action until the situation had been clarified and requested that Stanvac ask Shell to do likewise.[69] Walden readily agreed, and four more tankers rode at anchor.[70]

By August 4, State was ready to take another step. After asking Walden to hold matters in abeyance for at least one more week, Acheson called in Noel Hall, the British economic expert now attached to the Embassy in Washington, and Dutch Minister Dr. Loudon and Counselor Baron Van Boetzelaer.[71] In strict confidence Acheson gave them copies of the export quotas and restrictive specifications that were to be the unannounced guides for American policy. To give the Dutch time to decide their own policy, Acheson told Loudon that Stanvac and Shell had been asked to abstain temporarily from requesting export permits for their normal "quota" business and it was assumed that the Japanese would request permits for shipments under the Batavia contracts. The Dutch could do whatever they wished; if they issued export permits, the United States would free the necessary dollars for payment; if permits were denied, the United States would back the Dutch by refusing exchange licenses.[72] Presumably this polite passing of the buck derived from a basic American reluctance in mid-

[68]Memorandum of conversation, Hiss and Hornbeck with Walden and Singer, July 30, 1941, RG 59, 894.24/1566.

[69]Memorandum of conversation, Hornbeck with Walden, July 30, 1941, and memorandum, Hornbeck to Acheson and Welles, August 2, 1941, both in "Petroleum" file, Box 67, Hornbeck MSS.

[70]Telegram, Grew to SecState, September 5, 1941, FRUS(1941), V, 285.

[71]Memorandum of conversation, Acheson and Hiss with Hall, Loudon and Van Boetzelaer, August 4, 1941, FRUS(1941), V, 252–54,

[72]Ibid. Walden cabled his Yokohama office on August 5 not to take any action until further notice (memorandum, Hiss to Hornbeck and Acheson, August 5, 1941, RG 59, 811.20(D)Regulations/4034).

1941 to commit itself in advance to the defense of the Indies, for the overt use of influence on so sensitive an issue would have implied an attendant responsibility. The Dutch, however, were in a real dilemma since they were the exposed party and much preferred to keep in step with their strongest potential ally. For the moment they adopted an *ad hoc* policy, refusing export permits on various pretexts and admittedly watching the Japanese tankers still waiting in California for a clue to America's real intent.[73]

Behind this *ad hoc* position, another full embargo congealed. On August 5 Walden requested that the State Department provide Stanvac and Shell with copies of the specifications and quotas given to Hall and Loudon, and on August 14 Acheson assented. Regardless of what the Dutch finally decided, the two companies agreed that they would ship nothing from the Indies that would be outside the limits imposed in the United States.[74] The door was thus partly closed when Shell raised another problem. Both companies had traditionally operated their "quota" business in Japan under a form of extended credit, with exchange permits for conversion of yen to dollars or sterling running six or more months behind actual deliveries— which meant that both companies had considerable amounts in blocked yen accounts for products already delivered.[75] Shell's position was more extended than that of Stanvac, and it was Shell that suggested that regardless of export permits, no further quota business be shipped until these blocked accounts were paid out. Walden checked with the Department of State, received no objection, and decided to take a similar position.

[73]Memorandum of conversation, Hiss with Van Boetzelaer, September 15, 1941, *FRUS(1941)*, IV, 876–78.

[74]Memorandum, Hiss to Hornbeck and Acheson, August 5, 1941, RG 59, 894.24/1568; memorandum of conversation, Hiss with Walden, August 18, 1941, *FRUS(1941)*, V, 277–78.

[75]Memorandum by I. B. White on "Petroleum Distribution in Japan," July 15, 1940, RG 59, 894.6363/348; memorandum of conversation, Fox and Schmidt (Foreign Funds Control) with Walden and Singer (Stanvac), July 29, 1941, "Japan-Oil Shipments" folder, Box 513, RG 131.

The Japanese promptly agreed to the transfer of a $557,000 remittance coming due in August to New York, and by September 2 Stanvac had filed formal application with the Foreign Funds Control Committee for release of that amount from blocked Japanese accounts.[76]

The Dutch finally decided to handle their problem by the simple administrative device of requiring proof of an exchange license *before* issuing export permits, and since the original Batavia contracts had called for payment in dollars, this provision meant that these shipments also would hinge on action by the United States. Mitsui had already applied for a Dutch export permit to cover a cargo of Tarakan crude, and primarily for clarification of governmental intent Walden filed another application with Foreign Funds Control for release of $150,000 to cover payment from Mitsui.[77] No oil had left either the United States or the Indies for a month, and the tangle in both

[76]Memorandum, Hiss to Hornbeck, Hamilton, and Acheson, August 11, 1941, RG 59, 856D.6363/821–1/3; memorandum of conversation, Hiss with Walden, August 28, 1941, RG 59, 894.6363/387; "Excerpt of Memorandum for the Secretary's files for August 29th," "Japan-Oil Shipments" folder, Box 513, RG 131; minutes of Foreign Funds Control Interdepartmental Committee meeting of August 29, 1941, Vol. 436, pp. 173–75, Morgenthau Diary; memorandum of conversation, Hiss with Walden and Singer, September 2, 1941, and Hiss memorandum of September 11, 1941, both in "Petroleum" file, Box 67, Hornbeck MSS. This covered only the amount coming due in August; the total amount in Stanvac's blocked accounts was approximately $2 million.

[77]Letter, Walden to Hiss, August 12, 1941, RG 59, 894.24/1609; memorandum of conversation, Hiss with Walden and Singer, September 2, 1941, "Petroleum" file, Box 67, Hornbeck MSS; "Excerpt of Memorandum for the Secretary's Files of September 11," and resumé of telephone conversation between Fox (Foreign Funds Control) and Singer (Stanvac), September 17, 1941, both in "Japan-Oil Shipments" folder, Box 513, RG 131; memorandum, Miller (Foreign Funds Control) to Hiss, September 18, 1941, RG 59, 894.24/1738. As complicated as the narrative has now become, it is a simplification of the original episode, which included lengthy discussion of numerous extraneous difficulties in the transfer of funds. The British were frustrated bystanders as this drama unfolded (Foreign Office minutes of August 26 and 27, 1941, FO371(1941)F8504/1299/23).

cases now hinged on American willingness to release blocked funds. For the month of August at least, the administration of monetary controls created an unintended policy.[78]

Here was the point at which Hull asked that the complications be converted into a deliberate stall pending further "clarification." When Stanvac's two applications came before the Interdepartmental Committee on September 11, Acheson repeated the Secretary's request that the committee "continue to examine the problem" without either taking "any new restrictive measures" or relaxing its "present attitude."[79] The committee tabled Stanvac's applications without action, and the *de facto* embargo thereby spread to the Indies. Presumably as an inducement to break the deadlock, the Japanese government next offered to transfer $1,500,000 to Stanvac in settlement of its old Manchurian claim, if funds could be released from blocked accounts, but the Treasury simply added this to the list of applications on which no decision had been made.[80]

---

[78]Feis acknowledges this briefly in a footnote, but appears to have written without access to the records of Foreign Funds Control, which are much more explicit than State Department files on Hull's conversion of this tangle into deliberate policy (Feis, p. 261, note 1).

[79]"Excerpt of Memorandum for the Secretary's Files of September 11," "Japan-Oil Shipments" folder, Box 513, RG 131. This was the point at which the British finally obtained clarification of American intent. Quoting Acheson as his source, Sir R. I. Campbell cabled the Foreign Office on September 13 that "In the last two days Mr. Hull has given specific instructions that there is to be absolutely no weakening on economic front vis-à-vis Japan. Every device to delay the issue, including financial, must at all events be resorted to. At the same time there is to be no public (repeat public) decision or alteration in regulations which might demonstrate the completeness of the present embargo or suggest greater severity"; telegram, Campbell (Washington) to Foreign Office, September 14, 1941, FO371(1941)F9322/1299/23. According to American records, Acheson did not officially inform the British until September 27 (memorandum of conversation, Acheson with Hall, September 27, 1941, *FRUS(1941)*, IV, 887–88). It can be conjectured that Acheson decided to relieve British anxiety informally two weeks before he felt he could do so formally.

[80]Memorandum, Miller (Foreign Funds Control) to Hiss, September 24, 1941, RG 59, 840.51 Frozen Credits/3546–5/6.

The Dutch continued to worry over American intent, and by mid-September began to press hard for clarification, indicating that they were not averse to a complete embargo if that were the actual policy.[81] Acheson reviewed the situation in detail with Hamilton, Hornbeck, and Hull,[82] and on September 26 pointed out in strictest confidence to Van Boetzelaer that "through the medium of our freezing control, exports of petroleum to Japan have ceased and the Netherlands authorities may expect that through the same control the same result will continue."[83] On the following day the same oblique but diplomatically clear message was given to Hall of the British Embassy,[84] and on October 1 Acheson informed the Interdepartmental Committee that Hull "does not want any clear statement made on the question of oil exports to Japan, but rather prefers that a direct answer be delayed as long as possible."[85] For the next two months the Japanese continued to propose different methods of payment, but these advances apparently were designed solely to detect any subtle change in the American position. The Dutch adopted a procedure identical to that in the United States, the Batavia contracts expired quietly on November 1, and shortly thereafter the two Japanese tankers weighed anchor with no

[81]Letter, Troubeck (Ministry of Economic Warfare) to Ashley-Clarke (Foreign Office), September 15, 1941, FO371 (1941)F9415/1299/23; memorandum of conversation, Hiss with Van Boetzelaer, September 18, 1941, RG 59, 894.24/1753; and memorandum of conversation, Hiss with Van Boetzelaer, September 20, 1941, RG 59, 894.24/1754.

[82]Memorandum, Acheson to Hull, September 22, 1941, with attached memorandum by Hornbeck, September 24, 1941, *FRUS(1941)*, IV, 881–85; Acheson, p. 26.

[83]Memorandum of conversation, Acheson with Van Boetzelaer, September 26, 1941, *FRUS(1941)*, IV, 886–87.

[84]Memorandum of conversation, Acheson with Hall, September 27, 1941, *FRUS(1941)*, IV, 887–88; Medlicott, II, 117–18.

[85]Minutes of the Foreign Funds Control Interdepartmental Committee meeting of October 1, 1941, Vol. 447, p. 130, Morgenthau Diary.

cargo and left the West Coast.[86] Preplanned or not, the embargo was complete, and for all practical purposes had been so since the last cargo of crude left the Indies on August 5.

NOTHING in the accessible record explains Hull's change of position or Roosevelt's acquiescence, but at least three influences may be conjectured. In the first place, Hull was profoundly disillusioned by the fact that while he was negotiating with Nomura and doggedly trying to stave off sanctions, messages were intercepted that revealed Japan's aggressive intentions, and the Japanese moved into southern Indochina. The move destroyed hope that American restraint would have much effect on Japanese plans, and Hull returned to Washington August 4 convinced that the United States had "reached the end of possible appeasement with Japan and there is nothing further that can be done with that country except by a firm policy."[87] The emphasis now shifted to delaying an attack for as long as possible and buttressing American defenses.

Roosevelt appears to have been of like mind, for when pressed hard by Churchill at Argentia for a stern warning to

[86]Memorandum, Bernstein to White (Treasury), October 2, 1941, "Japan-Oil Shipments" folder, Box 513, RG 131; memorandum of conversation, Hull with Nomura, October 3, 1941, *FRUS(1941)*, IV, 891–92; memorandum of conversation, Fox with Nishiyama, October 10, 16, 17, 22, 24, 1941, and memorandum of conversation, Bernstein with Nishiyama, October 29, 1941, all in "Japan-Oil Shipments" folder, Box 513, RG 131; telegram, British Consul General at Batavia to Foreign Office, November 1, 1941, FO371(1941)F11704/1732/61; memorandum of conversation, Bernstein with Nishiyama, November 17, 1941, "Japan-Oil Shipments" folder, Box 513, RG 131, letter, J. S. Dent (British Embassy) to J. J. Reinstein (Department of State), November 21, 1941, RG 59, 894.24/1811; memorandum of conversation, Lawler (Treasury) with Desvernine (attorney for Mitsui), November 21, 1941, and memorandum of conversation, Fox with Kabacoff (attorney for Mitsui), November 25, 1941, both in "Japan-Oil Shipments" folder, Box 513, RG 131.

[87]Diary entry for August 8, 1941, Stimson MSS. See also Stimson's diary entries for August 7 and 12; Hull, II, 1015; Israel, p. 211; Feis, pp. 248–50; Langer and Gleason, *Undeclared War*, p. 659.

Japan, he came close to full agreement. But, as Churchill reported to Eden August 11, it was the "President's idea . . . to [continue to] negotiate . . . and thus procure a moratorium of, say, thirty days in which we may improve our position. . . . He will also maintain in full force the economic measures directed against Japan. These negotiations show little chance of succeeding, but [the] President considers that a month gained will be valuable."[88] The olive branch would continue to be extended, but hereafter it would be from a position of diminishing hope and increasing firmness. The Economic Defense Board was told in late August that "Japanese trade . . . policy . . . was a matter of confidential discussion between the President and Secretary Hull,"[89] but as late as August 28,

[88]Winston Churchill, *The Second World War* (6 vols.; Boston: Houghton Mifflin, 1948–53), III, 439. See also Welles' memoranda on the Argentia discussions, *FRUS(1941),* IV, 345–67; Sir Alexander Cadogan, *The Diaries of Sir Alexander Cadogan, 1938–1945,* edited by David Dilks (New York: Putnam, 1972), pp. 397–402; Robert E. Sherwood, *Roosevelt and Hopkins: An Intimate History* (New York: Harper, 1948), pp. 354–57; Charles A. Beard, *President Roosevelt and the Coming of War, 1941: A Study in Appearances and Realities* (New Haven: Yale University Press, 1948), pp. 454–61; Stanley W. Kirby, *The Loss of Singapore,* Vol. I in *History of the Second World War: The War Against Japan* (5 vols.; London: H. M. S. O., 1957), pp. 72–73; Theodore A. Wilson, *The First Summit: Roosevelt and Churchill at Placentia Bay 1941* (Boston: Houghton Mifflin, 1969), pp. 163–67; Feis pp. 255–56; Langer and Gleason, *Undeclared War,* pp. 670–77.

[89]Statement attributed to Acheson in a memorandum from Gaston to Morgenthau reporting the second meeting of the Board on August 20, 1941; Vol. 434, p. 211, Morgenthau Diary. The minutes of this meeting reveal that Wallace, the Chairman of the Board, thought crude was still going to Japan and suggested that it be cut off. Acheson said that Hull and Roosevelt were "working on this question and . . . it would prove embarassing if the Economic Defense Board went into the matter" (minutes of the Economic Defense Board meeting of August 20, 1941, "Economic Defense Board" folder, box 131, Robert P. Patterson MSS, Library of Congress, Washington, D.C.). Despite an extensive search, these were the only two documents located that even indirectly linked Roosevelt with Hull's September action, but it would be difficult to believe that he was not aware of it and at least acquiesced.

Roosevelt still believed there was nothing to prevent shipments "under the oil quotas allowed Japan."[90] It can be hypoth-
esized that when they learned in early September of the widespread belief that a full embargo had actually been imposed,
Roosevelt and Hull concluded that any change would be interpreted by Japan, Britain, and the American public as a sign of
weakness, and they decided to leave the situation as it stood.

Even if Hull had wanted to reopen the pipeline, he would
have had considerable difficulty. On September 15, the licensing personnel from his Controls Division had been transferred
along with Maxwell's Export Control staff to the new Economic
Defense Board, whose eight Cabinet-level members included
Wallace, Morgenthau, Stimson, and Knox—all advocates of
a tough line toward Japan.[91] Hull had lost the administrative
control used so effectively in 1940, and the best he could probably do was restrain Foreign Funds Control from a provocative
denial of the pending applications. In effect, the *de facto* embargo had been cast in administrative concrete by mid-
September, and no one appeared disposed to break it.

A full account of what transpired thereafter is well beyond
the scope of this study, but suffice it to say that in the final
analysis, oil proved to be the critical factor.[92] Unaware of all
the events just recounted, the Japanese watched actual tanker
movements, and from the first of August onward assumed

[90]Memorandum of conversation, Roosevelt and Hull with Nomura,
August 28, 1941, *FRUS(Japan 1931–41)*, II, 571–72.

[91]Executive Order No. 8900 and Administrative Order No. 1, both
September 15, 1941, cited in *Documents on American Foreign Relations*, IV, 718; Roosevelt, *Public Papers*, X, 293. A copy of the Executive Order is in "Economic Defense Board" folder, Item 150, Box 901,
RG 169.

[92]The final months before Pearl Harbor have generated violent controversy and hundreds of monographs. This brief interpretation is
synthesized from Feis, pp. 251–341; Langer and Gleason, *Undeclared
War*, pp. 693–731, 836–941; Butow, *Tojo*, pp. 228–403; Ike, pp. 122–
285; Iriye, *Across the Pacific*, pp. 200–226; Roberta Wohlstetter, *Pearl
Harbor: Warning and Decision* (Stanford: Stanford University Press,
1962); and Louis Morton, "Japan's Decision for War," in Kent R.
Greenfield, ed., *Command Decisions* (Washington, D.C.: Department
of the Army, 1960), pp. 99–124.

they had been completely cut off from American oil. They reacted just as most Asian oil experts had predicted for almost a decade—by deciding to seize the Indies. Although events of July 1941 had convinced many Americans that the Japanese were already committed to a military thrust southward, this was not the case. Policy discussion in Tokyo had clearly tended in that direction, but commitments (or for that matter plans) were seldom discussed very far in advance, and the embargo actually forced a decision that had not yet been made. With oil reserves for only eighteen months of wartime operations, the clock began to tick the first of August, and a mounting sense of desperation permeated Tokyo. As the Japanese perceived the options, they could do nothing and let their war-making capacity drain away; they could give up their hard won gains on the mainland to gain respite from Anglo-American sanctions; or they could make a final desperate lunge for autonomy. They chose the last with a plan to protect their flanks by disabling the American fleet at Pearl Harbor, thrusting south to gain and hold the resources of Southeast Asia, and gambling on a German victory or a failure of American will to avoid a long test of strength with the United States. It was a reckless way out of a situation they themselves had created, but it might have worked. Stanvac played one final though unsolicited role in the drama. Japan's primary objective on December 7 was the oil of Royal Dutch-Shell and the Standard-Vacuum Oil Company in the Netherlands East Indies. The battleships at Pearl Harbor were destroyed primarily to protect the long tanker route from Sumatra to Honshu.

EPILOGUE AND
CONCLUSIONS

THE SHOCK OF the Japanese attack on Pearl Harbor diverted the attention of the general public from Japanese strategy in the opening days of the Pacific War. Americans were conscious of Japanese advances in the Philippines, Thailand, Malaya, Sumatra, and Java, but the loss of American lives and ships at Pearl Harbor proved so traumatic that the fate of the Indies oil fields received only passing notice amid the unrelenting gloom of early 1942. Japanese troops had begun landing at Kota Bharu on the Malay Peninsula two hours before the strike at Hawaii,[1] and the Malay campaign moved so swiftly that by February 15 the British bastion at Singapore fell to a numerically inferior Japanese force. The invasion of the Netherlands East Indies began January 11 and for all practical purposes was completed by the end of March 1942. The huge Soengi Gerong refinery, which Walden had worked so hard to build at Palembang, Sumatra, went up in flames just ahead of its capture by the Japanese on February 15, by grim coincidence the same day that Singapore fell. Japanese paratroopers attempted to seize the refinery intact, but landed slightly off target, and were held off long enough by a small contingent of Dutch and British troops for the prearranged demolition plans to be executed. Wells in the Talang field were sealed with cement, the connecting pipelines were destroyed, and the sixteen remaining Americans, led by NKPM managing director Elliott, escaped by small craft to Java and thence Australia.[2] Similar destruc-

---

[1]John Toland, *The Rising Sun: The Decline and Fall of the Japanese Empire, 1936–1945* (New York: Random House, 1970), pp. 204, 230.

[2]Letter, W. J. Gallman (Department of State) to Walden, February 20, 1942, Exxon files; letter, Elliott (then in Melbourne) to Walden, March 29, 1942, Exxon files; "The Fiery End of Soengi Gerong," *The Lamp,* April 1942, pp. 24–27, Exxon library.

193

tion plans were carried out with varying degrees of success on Stanvac and Shell facilities throughout the Indies, but by the end of March the long coveted territory was completely within Japanese control.

Drilling crews and oil-field technicians landed in the Indies with the invading forces, and by the end of 1942, the Japanese had brought in four thousand trained personnel, close to seventy percent of the total available from the home islands.[3] Within two months these crews had commenced production from newly-drilled wells and some cleaned-out old ones, and by 1943 had restored output to nearly 50 million barrels of crude, approximately three-fourths the 65 million barrels produced in 1940.[4] Unlike Soengi Gerong, the Shell Pladjoe refinery outside Palembang had been captured almost intact by paratroop attack and was back in operation within three months, with several other refineries restored to at least partial operation by September. Refinery output in 1943 reached 28 million barrels, close to half the 64 million barrels produced in 1940,[5] but by that time the nightmare feared by Japanese naval strategists had begun to materialize. The long tanker route back to the home islands came under increasing submarine, air, and mining attack, sharply reducing the quantity of oil actually reaching the home islands. By late 1944 this lack forced drastic reductions in Japanese air and naval operations, and when the last tanker convoy leaving Singapore in March 1945 failed to reach its destination, the situation had become desperate. Domestic

[3]This account of Japanese exploitation of Indies oil is based on Cohen, pp. 140–47; and U. S. Strategic Bombing Survey, *Oil in Japan's War*, Final Reports 51 and 52 (Statistical Appendix), Report of the Oil and Chemical Division, February 1946, Record Group 243, Records of the U. S. Strategic Bombing Survey, National Archives, Washington, D.C., hereafter cited as "RG 243".

[4]Cohen, p. 140. This data is approximate since the detailed Japanese records accumulated at Singapore were destroyed at the end of the war, and Cohen relied on estimates made in Tokyo. The figures presumably include some crude produced in British Borneo as well as the Indies.

[5]U. S. Strategic Bombing Survey, Report No. 51, pp. 49–50, RG 243.

production remained minimal, and the synthetics on which Japan had placed so much reliance failed to materialize.[6] Japan began the war with a reserve of 43 million barrels, but by July 1945 total inventories had dwindled to 3 million barrels with domestic production remaining inadequate even to replace that amount. The Empire had not been spared the long test of strength with the United States, thus losing its only hope of winning the war, and its ability to continue modern air and naval warfare was drastically curtailed even before the introduction of nuclear weapons at Hiroshima and Nagasaki. Though many other factors intervened, the course of the Pacific War was in large measure determined by Japan's struggle and ultimate failure to secure for itself a flow of military fuel from the Netherlands East Indies. The gamble had been taken and lost.

AND SO WE have a chronicle of events as they appear to one observer examining traces in the sand of surviving documentation through a particular set of perceptual filters. If nothing else, it should be abundantly clear that there is no simple answer to the original question of the degree and direction of Stanvac's influence on American Asian policy prior to Pearl Harbor. The relationship was intricate, evolutionary, and parallel to other influences, which must also be weighed in order to evaluate Stanvac's role in perspective. But men have penchants for organizing random events into comprehensible patterns in order to deal with an elusive reality, and with full realization that others may see different patterns, it appears useful to make a broad observation on the period before attempting a precise analysis of the role played by Stanvac.

Earlier in this study it was noted that the Anglo-American oil business in East Asia was in the process of rationalization in the 1930s, even before it began to feel the pressure of Japan's drive for autonomy. It can be argued that Japanese

[6]Total synthetic oil production, much of it in Manchuria, reached 1,048,000 barrels in 1943, less than eight percent of the 14,046,000 originally projected for that year (Cohen, p. 137).

expansion simply accelerated this process as a tightly knit informal organization developed to counter the threat, but that statement requires explanation. As used here the term "rationalization" simply means the organization of activity into well defined and well coordinated packages under hierarchical direction to reduce the risks of unpredictable behavior and to increase the efficiency of the overall operation—whatever it may be. Eleven men on a grass field are "rationalized" when they are organized into a football team with assigned positions and defined responsibilities each time the quarterback calls specific signals. If that play happens to be performed solely as a scene in a motion picture and the director gives both teams predetermined instructions to insure a predictable outcome (such as a scoring run around left end), the process of rationalization has gone one step further, and the teams are actually performing as coordinated units toward a common objective—in this case a specific dramatic sequence. Transferring this analogy to the oil business, the initial combination of production, transportation, refining, and distribution into vertically integrated companies represented an early stage in the rationalization process. And as has been already noted, the formation of Royal Dutch-Shell, the Red Line Agreement in the Middle East, and the merger that produced Stanvac itself all represented further steps toward rationalization of large segments of the industry.

If rationalization may be applied to informal as well as formal organization, it is quite apparent that Japanese expansion accelerated this process in East Asia by prompting development of an increasingly well coordinated team among those institutions most concerned over its drive for autonomy in oil—Stanvac, Shell, the Department of State, and the Foreign Office. Although similar institutional interests created and sustained the bond, these interests were not identical and they changed with time. In the early 1930s Stanvac and Shell were most concerned with protecting their marketing outlets in Japan, Manchuria, and China, and diplomats chafed principally over violations of Open Door treaty rights. As the decade progressed the two companies, and Stanvac in particular, shifted concern to the protection of their source of supply and heavy investment in the Indies, while the two governments, and the United States in particular, began to link their own security with defense (or at least avoidance of conflict) in the same area. Again, institutional interests intersected and the teamwork continued. At least until July 1941, control of policy was increasingly centralized in the Department of State, which held the key to the ultimate weapon—an oil embargo. By 1940 and 1941 more and more agencies were becoming involved in Asian oil, and the complex administrative network became increasingly susceptible to generalized attitudes rather than unilateral direction. Behavior within this whole interlocked framework more nearly resembled the complex power struggles described by Crozier in *The Bureaucratic Phenomena* than a neat interchange of institutional views on isolated policy questions.

The point is to cast doubt on any simple analysis of "influence." If in fact Stanvac operated increasingly as one segment of an integrated team, the point of reference should really be the lines of actual control and the informal politics within the entire administrative structure even though technically this informal structure might cross institutional boundaries. Under stress, lines of demarcation between diplomatic, military, and corporate responsibility became increasingly blurred. Stated more simply, the fact that Stanvac (or anyone else) took a given position in a letter to the Treasury might be completely irrelevant if the line of true administrative control lay elsewhere; conversely, obscure action by another Stanvac official might actually produce the opposite result. For these reasons a *point by point* analysis appears more appropriate than any generalization covering the entire period.

In this context, it is evident that an embryonic pattern of cooperation in East Asia existed at least as early as the Canton kerosene war of 1933 and 1934, when oil men and diplomats worked closely as long as both markets *and* treaty rights were threatened. This cooperation was essentially on a local *ad hoc* basis, and it would be difficult to argue that Stanvac in that instance had any significant impact on general American policy or relations with China—particularly since it fit within a long established pattern of diplomatic support for American business in East Asia whenever treaty rights came under attack. The episode did shed light, however, on the different sources of corporate and diplomatic motivation that appeared in reactions to the Manchurian oil monopoly and Japanese Petroleum Industry Law of 1934. Stanvac and Shell concentrated on defense of their far larger market in Japan while Anglo-American diplomats reacted strongly to violations of the Open Door policy in Manchuria. The two issues intertwined, the common opponent was Japan, and knowledgeable teamwork began to mature in New York, Washington, and London. The real issue was to protect Stanvac and Shell's distribution system in Japan without spending company funds to support Japanese stockpiling, and Stanvac's Parker emerged as the key strategist in that campaign. In retrospect the exodus from Manchuria and the carefully cultivated *hint* of a worldwide embargo on crude oil were principally for psychological effect, with the embargo idea taken seriously only within the confines of the British Foreign Office. As events turned out, Parker's extended campaign proved successful. Stanvac and Shell were able to enlist enough diplomatic support to continue operations right down to 1941 without ever participating in Japanese stockpiling. How much effect this episode had on Japanese-American relations is problematic, but it fit within a general pattern of American irritation over Japanese behavior and an increasing Japanese sense of isolation in a hostile world. Neither side was seriously hurt, but the episode added one more source of friction in an already deteriorating situation.

The Sino-Japanese war, which began in 1937, had less effect on Stanvac and Shell than might have been expected, since it only extended the problems and defensive tactics already developed in Japan to the China mainland. The companies were severely handicapped in the Yangtze Valley, but they staved off a monopoly in Inner Mongolia and held their own in North China. Stanvac received prompt compensation for the three small tankers sunk along with the *Panay,* and the companies also succeeded in maintaining a substantial China trade down to 1941. Again it would be difficult to argue that Stanvac uniquely influenced American policy, but the experience compounded a growing sense of diplomatic frustration over Japanese treaty violations in China. In this perspective, Stanvac's role through 1938 could probably be summarized as follows: because of its exposed position the company did initiate requests for diplomatic support against the effects of Japan's drive for autonomy and to this extent provided additional grist for the mill of deteriorating relationships.

After 1938 Walden became increasingly concerned that the oil fields of the Netherlands Indies would be Japan's ultimate target, and corporate emphasis shifted to support of that institution most likely to provide an effective defense—the government of the United States. As has been noted, a major factor in this shift may have been Walden's concern for an operation he had personally built up, but it was not inconsistent with the fact that the Indies represented Stanvac's major source of supply and by far its largest investment in East Asia. Whatever the reason, this change in emphasis coupled with increasing American governmental concern over the security of the same area produced an exceptionally close working relationship between Stanvac and those agencies most concerned with Asian oil– the State Department, the Navy, Export Control, and the Treasury. In 1940 when Roosevelt and Hull fought off embargo advocates to avoid (or at least delay) a confrontation with Japan in the Indies, it was Walden and Shell's Agnew who worked out a temporizing settlement providing Japan with some Indies oil while Hull quietly used his administrative veto to negate even the partial embargo on aviation gasoline imposed on Japan. Stanvac's role in this period could only be characterized as an implementor of presidential policy, and the real check on Japanese procurement proved to be a shipping shortage and a ban on metal drums—the latter a measure actually suggested by Walden and Parker. The idea was to avoid both a confrontation and excessive stockpiling.

By mid-1941, however, when the critical decision to curtail Japan's oil supply was finally in the offing, control had begun to slip from the hands of the original team. When Japan's occupation of southern Indochina convinced many that further temporizing was pointless, the team and the policy of restraint disintegrated simultaneously. Despite Roosevelt's decision to reduce rather than terminate Japan's oil supply, the freezing order threw control to an already hostile Treasury Department, and in the following month of confusion no oil was shipped from either the United States or the Indies. During August 1941 administration made policy, and Stanvac was a full participant in the Indies phase of this tangle. Since Stanvac was actually responding to requests of the Assistant Secretary of State and since Hull subsequently ratified the *de facto* embargo, it would be difficult to characterize Stanvac as anything other than an implementor in this phase also. Certainly the company suggested no new departures. The fact that this *de facto* embargo unwittingly started the clock ticking on Japan's final decision forward was no one's fault and everyone's fault—it would be difficult to isolate a single culprit in the atmosphere of late 1941. Both Japan and the United States were responding to what they perceived as a threat to their own security. In its own sphere of operations, the Standard-Vacuum Oil Company had been a full participant on the American side of this process.

APPENDIX A

STANVAC'S INVESTMENT IN EAST
AND SOUTHEAST ASIA

A FULL assessment of Stanvac's role in American Asian policy prior to World War II requires an estimate of the amount and distribution of capital the company had invested in the area threatened and later occupied by Japan. Information on this topic is available but must be treated with caution because of the inherent difficulty of interpretation. The "value" of an investment may legitimately mean cost, book, or estimated market value; real estate values fluctuate widely; methods of computing depreciation vary; and exchange rates seldom remain constant. Despite these hazards, an attempt has been made to arrive at a reasonable approximation of Stanvac's investment position in East and Southeast Asia, with the overall results shown in Table A–1.

All of the figures except those for the Netherlands East Indies were taken from Stanvac's response to a 1943 Treasury survey of prewar foreign investment[1] and were found to be consistent with information submitted in support of war damage claims after World War II.[2] The figures for the Netherlands East Indies were derived from data provided by the company

[1]Form TFR–500: Census of Property in Foreign Countries, submitted March 27, 1944, by the Standard-Vacuum Oil Company, Record Group 265, Records of the Foreign Funds Control, National Archives Branch, Suitland, Maryland, hereafter cited as "Stanvac TFR–500, RG 265."

[2]Esso Standard Eastern Claims for World War II Damages in China, Burma, French Indochina, Malaya, and the Philippines, submitted to the Foreign Claims Settlement Commission of the United States, January 14, 1965, with revisions through September 19, 1966, copies in the custody of the controller, Esso Eastern, Inc., New York, hereafter cited as "Esso Eastern War Damage Claim, (country)."

TABLE A — 1

ESTIMATED DIRECT INVESTMENT OF THE
STANDARD-VACUUM OIL COMPANY
IN EAST AND SOUTHEAST ASIA, 1941
(In thousands of dollars[a])

East and Southeast Asia (excluding N.E.I.)	Value
Japan (including Korea)	7,748
China (including Manchuria)	8,943
Hong Kong	3,115
French Indochina	1,137
Philippine Islands	6,673
Thailand (Siam)	52
Burma	464
British Malaya, Singapore, and Borneo	3,563
Subtotal	31,695
Netherlands East Indies	
Producing, refining, and marketing operations	74,405
Estimated value of petroleum reserves	150,850
Total value of investment[b]	256,950

[a]Gross book value in all cases except N.E.I., which is presumably cumulative capital expenditures; the value of petroleum reserves is a contemporary estimate only and does not represent actual costs incurred; see accompanying text for sources and explanation.

[b]Not including an estimated $6.6 million in fixed assets of the subsidiary Oriental Trade and Transport Company, Ltd., which operated a small tanker fleet under British registry out of Hong Kong.

to the Department of State in 1940,[3] and were checked against a number of other sources for reasonableness. The overall estimate was compared with published results of the 1943 Treasury survey,[4] contemporary investment studies of Cleona

[3]Letter, Walden to Hornbeck, September 16, 1940, RG 59, 756D. 94/84.

[4]U. S. Treasury Department, *Census of American-Owned Assets in Foreign Countries* [*1943*] (Washington, D.C.: Government Printing Office, 1947).

Lewis, Carl Remer, and Helmut Callis,[5] and a prewar Department of Commerce series on foreign investment[6] to reach the conclusion that Stanvac did, in fact, represent the single largest American direct investment in East and Southeast Asia immediately prior to Pearl Harbor. The remainder of this Appendix is essentially substantiating detail to the foregoing statements.

Corporate Structure

For a variety of reasons international oil companies have frequently elected to organize and reorganize their activities through legally incorporated subsidiaries rather than through simple divisions of the parent company. This practice spawned a multitude of corporate names which sometimes made it difficult to unravel the interests involved. For the sake of clarity, an abbreviated description of Stanvac's corporate structure in the 1930s is presented before going on to an analysis of its investment.

The Standard-Vacuum Oil Company was incorporated September 7, 1933, under the laws of the State of Delaware,

[5]Cleona Lewis, *America's Stake in International Investments* (Washington, D.C.: Brookings Institution, 1938); Carl F. Remer, *Foreign Investments in China* (New York: Macmillan, 1933); Helmut G. Callis, *Foreign Capital in Southeast Asia* (New York: Institute of Pacific Relations, 1942).

[6]U. S. Department of Commerce, Bureau of Foreign and Domestic Commerce, *American Direct Investments in Foreign Countries— 1929,* Trade Information Bulletin No. 731, by Paul D. Dickens (Washington, D.C.: Government Printing Office, 1930); *A New Estimate of American Investments Abroad,* Trade Information Bulletin No. 767, by Paul D. Dickens (Washington, D.C.: Government Printing Office, 1931); *American Direct Investments in Foreign Countries —1936,* Economic Series No. 1, by Paul D. Dickens (Washington, D.C.: Government Printing Office, 1938); and *American Direct Investments in Foreign Countries—1940,* Economic Series No. 20, by Robert L. Sammons and Milton Abelson (Washington, D.C.: Government Printing Office, 1942). These are hereafter cited as *"American Direct Investments—(year)."*

with main offices at 26 Broadway, New York (later moved to White Plains, New York). It represented a consolidation of the Asian producing and refining operations of the Standard Oil Company (New Jersey) with the Asian, South and East African, New Zealand, and Australian marketing operations of the Socony-Vacuum Oil Company, and those two parent corporations each retained a fifty-percent interest in the new joint subsidiary.[7] Socony-Vacuum itself was the product of a 1931 merger of the Standard Oil Company of New York with the Vacuum Oil Company, both former components of the Standard Oil group dissolved by the Supreme Court in 1911. Socony-Vacuum changed its name to Socony Mobil Oil Company in 1955 and then to Mobil Oil Corporation in 1966; the Standard Oil Company (New Jersey) changed its name to Exxon Corporation in 1972.[8] The Standard-Vacuum Oil Company was itself reorganized out of existence for a combination of business and legal reasons on March 30, 1962, with most assets divided between the two parent corporations. Legal continuity was provided by changing the name of Stanvac to Esso Standard Eastern, Inc. and continuing this corporation as a wholly owned subsidiary of the Standard Oil Company (New Jersey).[9] All of this may be condensed by simply noting that from 1933 until 1962 the Standard-Vacuum Oil Company (later Esso Eastern) was a jointly owned subsidiary of the Standard Oil Company (New Jersey), later Exxon, and Socony-Vacuum, later Mobil.

Stanvac's territory ranged from South Africa to Japan, Australia, and the Philippines, and in many cases operations

[7]*Moody's Industrial Manual,* 1939, p. 3088.

[8]*Ibid.,* 1970, p. 2655; 1973, p. 754.

[9]The reorganization appears to have been prompted by a growing divergence of business interests between Jersey and Mobil, but the plan also made possible a 1960 consent decree settling the Jersey portion of antitrust litigation pending since 1953. A few properties continued to be jointly owned after 1962, including P. T. Stanvac Indonesia: *Moody's Industrial Manual,* 1970, pp. 2272, 2655; *The Lamp,* Spring 1961, pp. 2–3, Exxon library; *Platt's Oilgram,* November 16, 1960, pp. 1–4, Exxon library.

were actually conducted by its own subsidiaries. The Vacuum
Oil Company of South Africa, Ltd., with headquarters in
Capetown, was responsible for sales in South and East Africa,
and the Vacuum Oil Company, Pty., Ltd., in Melbourne, sold
throughout Australia and New Zealand. Marketing in India,
Burma, and Ceylon was handled by an office of the Standard-
Vacuum Oil Company itself, located in Calcutta.[10] In 1936
the Standard-Vacuum Oil Company, Japan Division, with
offices in Yokohama, was responsible for marketing operations
in Japan, Formosa, Korea, Manchukuo, and the Japanese
portion of Sakhalin Island. The Standard-Vacuum Oil
Company, North China Division, based in Shanghai, was assigned a territory including China north of Foochow, Mongolia,
Sinkiang, and Tibet. The Standard-Vacuum Oil Company,
South China Division, with offices in Hong Kong, handled
marketing in China south of Foochow, French Indochina, Siam
(later Thailand), the Federated Malay States, Straits Settlement, Macao, British North Borneo, and the Philippine Islands.[11] Also based in Hong Kong was the subsidiary Oriental
Trade and Transport Company, Ltd., operating a fleet of ten
tankers under British flag prior to the outbreak of war.[12]

Four separate subsidiaries, all incorporated under Dutch
law, functioned in the Netherlands East Indies. The largest and
oldest, N. V. Nederlandsche Koloniale Petroleum Mattschappij
(NKPM), had been organized April 24, 1912, by Jersey to
handle exploration, producing, and refining in the Indies and
continued this activity as a Stanvac subsidiary. NKPM itself
held a forty-percent interest in N. V. Nederlandsche Niew

[10]*Socony-Vacuum News*, December, 1936, pp. 1–7, Mobil library.
[11]Ibid.

[12]Larson, Knowlton, and Popple, pp. 207, 394–95; *Socony-Vacuum
News*, September 1935, p. 13, Mobil library; Stanvac TFR–500,
RG 265; East Asia Research Institute, II, 596–97. The four sources
cited are somewhat contradictory in detail; Stanvac operated a sizeable
tanker fleet, but its organization and registry changed frequently. Fixed
assets of the Oriental Trade and Transport Company are given as
$6.6 million on the consolidated balance sheet of December 31, 1942
(attachment to Stanvac TFR–500, RG 265).

Guinee Petroleum Maatschappij (NNGPM), incorporated May 9, 1935, for exploration in New Guinea. Partners in this venture along with NKPM were subsidiaries of Royal Dutch-Shell (forty percent) and the Standard Oil Company of California and the Texas Company (which through *their* joint subsidiary, Caltex, held the remaining twenty percent). Marketing in the Indies fell under the jurisdiction of the N. V. Koloniale Petroleum Verkoop Maatschappij (KPVM), with a small tanker fleet sailing under the control of the N. V. Nederlandsche Koloniale Tankvaart Maatschappij (NKTM).[13] For simplicity all of these components are collectively referred to as "Stanvac" unless the context requires more precise identification.

Marketing Operations in East and Southeast Asia (Excluding N.E.I.)

As a fully integrated oil company, Stanvac complemented its producing and refining operations in the Netherlands East Indies with a vast marketing network throughout East and Southeast Asia, largely inherited from the old Standard Oil Company of New York. Although concentrated in China, distribution facilities were spread throughout the region; they included storage tanks, godowns (warehouses), service stations, deepwater and river terminals, trucks, railroad cars, barges, and oceangoing tankers.[14]

[13]Esso Eastern War Damage Claims, Netherlands East Indies; Alex L. Ter Braake, *Mining in the Netherlands East Indies* (New York: Institute of Pacific Relations, 1944), pp. 68–69; *Socony-Vacuum News*, July, 1935, p. 7, Mobil library; *Moody's Industrial Manual*, 1971, pp. 1376, 2514. In 1947 NKPM changed its name to Standard-Vacuum Petroleum Maatschappij (SVPM), and KPVM adopted the designation Standard-Vacuum Sales Company (SVSC), with a further name change to P. T. Stanvac Indonesia (PTSI) in 1959. Under the 1962 agreement SVPM and PTSI were consolidated into P. T. Stanvac Indonesia and continued as a joint subsidiary of Jersey and Mobil. In 1962 NNGPM became Sorong Petroleum Maatschappij, N. V. (SPM), and in 1964 this corporation was sold to the Indonesian government.

[14]Esso Eastern War Damage Claims, China, Burma, French Indochina, Malaya, the Philippines, and Netherlands East Indies.

The marketing territory of the Japan Division was aligned with the area under Japanese political control and in the 1930s included Korea and Formosa. Headquarters were in Yokohama, with primary storage installations at Tsurumi (above Yokohama), Nagasaki, Shimonoseki, Kobe, and Otaru. Small storage tanks and warehouses for packaged products were located in most of the other large cities of Japan.[15] A similar network existed in Korea, with the major storage installation at Keijo (Seoul), and much more limited facilities were located in Formosa.[16]

The oldest and most extensive marketing network in the Orient was that developed in China. Branches were located in most major Chinese cities, including Peiping, Chinwangtao, Tientsin, Chefoo, Tsingtao, Tsinan, Chungking, Hankow, Kiukiang, Nanking, Changsha, Canton, Foochow, Amoy, Swatow, and Yunnanfu, with numerous suboffices deep in the interior.[17] The North China Division also operated a fleet of small tankers, ranging from 274 to 1,118 gross tons, up the Yangtze River to Hankow, Ichang, and Chungking, and along the coast to Ningpo and Wenchow.[18] The territory of the South China Division extended to Thailand, French Indochina, British Malaya, Singapore, North Borneo, and the Philippines, with facilities similar to, but not as extensive as those in China. Principal installations were located at Saigon, Singapore, and

[15]Naval Intelligence Report, DIO 1ND Serial 83–43, November 3, 1943, Report No. 62203R, Record Group 226, Records of the Office of Strategic Services, Research and Analysis Branch, National Archives, Washington, hereafter cited as "RG 226."

[16]Attachment dated May 1934, to memorandum of conversation, Hornbeck, Phillips, Teagle, and Deterding, August 22, 1934, RG 59, 894.6363/84.

[17]East Asia Research Institute, I, 545; Letter, Johnson (Minister to China) to SecState, April 10, 1934 ("Report on competition in the China market for petroleum and its products"), RG 59, 893.6363/110.

[18]East Asia Research Institute, II, 590–91. Thirteen of these small tankers were reportedly operational in 1937, but the three largest of these, the *Mei Ping, Mei Hsia,* and *Mei An,* were destroyed by Japanese aircraft along with the *U.S.S. Panay* in December of that year; see telegram, Hull to Grew, April 7, 1938, *FRUS(Japan, 1931–41)* I, 561–62.

TABLE A — 2

GROSS BOOK VALUE OF STANVAC ASSETS
WRITTEN OFF AS WAR LOSSES

Country	Value
Japan	$ 7,747,642
China (Manchukuo)	115,308
China (Occupied)	8,828,093
Hong Kong	3,114,671
French Indochina	1,137,373
Philippine Islands	6,672,622
British Malaya	3,562,733
Thailand (Siam)	51,634
Burma	463,689
Total	$31,693,765

SOURCE: Note A to Stanvac TFR–500, RG 265.

Manila.[19] As noted earlier, this network was rounded out by the tanker fleet of the Oriental Trade and Transport Company based in Hong Kong.

Realistic valuation of such a far-flung and long-standing marketing network is exceedingly difficult, but primary reliance has been placed on a set of figures reported as an addendum to Stanvac's response in the 1943 Treasury survey as the "... gross book value of ... [assets] ... written off as war losses ... in Japan ... and in certain other countries occupied by Japan ..." on its consolidated balance sheet for December 31, 1942, as shown in Table A–2.[20]

These figures proved to be identical with those given as prewar "gross assets" before depreciation in substantiation of war damage claims submitted in 1965 for China (including

[19]Esso Eastern War Damage Claims, French Indochina, Malaya, and the Philippines; Schedule No. 1 attached to Stanvac TFR–500, RG 265.

[20]Note A to Stanvac TFR–500, RG 265.

Manchuria and Hong Kong), French Indochina, the Philippines, and Malaya (including Singapore and North Borneo) and Burma.[21] The actual claims for war damages were not directly comparable to prewar investment since they did not include the value of property recovered, did include rehabilitation expenses, and were based on different depreciation rates than those used for book purposes. The net war damage claims, however, were either reasonably close to the values noted as prewar "gross assets" or differed for logical reasons (such as claims already paid by another government). No claim was submitted for Japan since under the final peace treaty all assets were returned with damages repaired by the Japanese government.[22] The figure of $7.7 million in "gross book value" for Japan is very close, however, to a value of $7.1 million for "first cost" of fixed assets in Japan, Korea, and Formosa given to the Department of State by Jersey president Teagle in 1934.[23] One other set of figures, the ones actually submitted by Stanvac as valuations for the 1943 Treasury survey, were somewhat lower than the ones given above, and appear to have been heavily depreciated values.[24] Considering all of this, $31.7 million seems a reasonable approximation of Stanvac's 1941 investment in East and Southeast Asia, exclusive of the Netherlands East Indies.

Netherlands East Indies

Associated with this marketing network were the production and refining operations of the Nederlandsche Koloniale Petro-

[21]Esso Eastern War Damage Claims, China, French Indochina, the Philippines, Malaya, and Burma.

[22]Letter to the author from Mr. L. G. Belcher, controller, Esso Eastern, Inc., New York, March 3, 1971.

[23]Attachment dated May 1934, to memorandum of conversation, Hornbeck, Phillips, Teagle, and Deterding, August 22, 1934, RG 59, 894.6363/84. This was a complete summary of investments in Japan presented in support of Teagle's request for State Department assistance in protesting the Japanese Petroleum Industry Law.

[24]Stanvac TFR–500, RG 265.

leum Maatschappij (NKPM) in the Netherlands East Indies. NKPM's major installation in 1941 was a 46,000 barrel/day refinery at Soengi Gerong, ten miles south of Palembang in Southern Sumatra.[25] By 1940 this refinery was equipped to produce 100-octane aviation gasoline as well as motor gasoline, kerosene, diesel, and fuel oil, and could more than handle the full production of NKPM's prolific Talang Akar field eighty miles to the southwest.[26] While Talang Akar accounted for the lion's share of NKPM's Indies production, the company also conducted exploration and production operations in northern and central Sumatra, east Java, and east Borneo, with a total of 529 producing wells in 1940 and a prewar peak output of 43,350 barrels/day in 1939.[27] The company owned office buildings in Batavia, a small 500 barrel/day refinery at Kapoean in central Java, and a major ocean terminal at Tandjung Uban on the island of Poelau Bintan forty miles southeast of Singapore.[28] As already noted, NKPM also held a forty-percent interest in Nederlandsche Niew Guinee Petroleum Maatschappij (NNGPM), organized jointly in 1935 with subsidiaries of Royal Dutch-Shell, Standard Oil of California, and The Texas Company for exploration in New Guinea. In addition to the normal equipment, NNGPM owned and op-

[25]Esso Eastern War Damage Claim, Netherlands East Indies; *The Lamp*, April 1942, pp. 24–27, Exxon library; Summary of Standard-Vacuum Oil Company 1938–39 Operations, prepared October 2, 1946, Exxon files.

[26]Esso Eastern War Damage Claims, Netherlands East Indies; *SVPM, Standard-Vacuum Petroleum Maatschappij: Forty Years of Progress, 1912–1952* (Djakarta, Indonesia: Standard-Vacuum Petroleum Maatschappij, 1953), pp. 26–27, Exxon library; *Socony-Vacuum News*, February, 1935, p. 7, Mobil library.

[27]Ter Braake, pp. 68–71; Letter, G. S. Walden (Stanvac board chairman) to W. C. Teagle (Jersey board chairman), April 4, 1940, Exxon files; Summary of Standard-Vacuum Oil Company 1938–39 Operations, prepared October 2, 1946, Exxon files; *SVPM: Forty Years of Progress*, p. 21, Exxon library.

[28]Esso Eastern War Damage Claims, Netherlands East Indies; Summary of Standard-Vacuum Oil Company 1938–39 Operations, prepared October 2, 1946, Exxon files.

erated an airfield and a number of aircraft for aerial mapping.[29] While NKPM production and refining accounted for most of its Indies investment, Stanvac also operated a marketing network similar to that in the rest of Asia through Koloniale Petroleum Verkoop Maatschappij (KPVM) and a small tanker fleet through Nederlandsche Koloniale Tankvaart Maatschappij (NKTM).[30] In short, Stanvac's investment in the Netherlands East Indies was extensive and diverse.

A curious complication arises in attempting a realistic valuation of this investment. Instead of including the assets of its Indies subsidiaries directly in its consolidated balance sheets, Stanvac carried NKPM, KPVM, and NKTM on its financial records at the nominal book value of its stock in those companies, and this appears to have had no relation to the actual cost of facilities in the Indies.[31] When Stanvac responded to the 1943 Treasury survey, it submitted a figure for the Indies that appears to have been the book value of its investment in these companies plus an amount for intercompany accounts receivable.[32] Technically this represented its "investment" in the Indies. The figure ($20.7 million) appears much too low for the actual value of its Indies facilities and is inconsistent with all other estimates. On the other hand, the postwar claim for war damages includes a sizeable amount for crude oil expropriated from NKPM fields by the Japanese during the war, and this produced a total claim ($97.5 million)[33] that appears much higher than actual prewar investment in facilities. Between these two lies a figure of $74,405,000, which

[29]Esso Eastern War Damage Claims, Netherlands East Indies; Ter Braake, p. 69.

[30]Esso Eastern War Damage Claims, Netherlands East Indies. In 1935, this fleet was reported to consist of four vessels, but by late 1941 these had been transferred to the control of NKPM under Belgian registry; *Socony-Vacuum News,* July 1935, p. 7, Mobil library; Larson, Knowlton, and Popple, pp. 207, 394–95.

[31]Standard-Vacuum Oil Company Consolidated Balance Sheet for December 31, 1942, attached to Stanvac TFR–500, RG 265.

[32]Stanvac TFR–500, RG 265.

[33]Esso Eastern War Damage Claims, Netherlands East Indies.

APPENDIX A

Stanvac board chairman Walden gave the Department of State as the company's "total investment in physical properties" in the Netherlands East Indies in 1940.[34] In another context in the same year, Walden wrote Jersey board chairman Teagle that "the total capital expenditures by NKPM for producing and refining assets during the twenty-eight years of its existence amounted to $71,217,000."[35] This suggests that the figure of $74.4 million given to the Department of State represented cumulative capital expenditure of NKPM, KPVM, and NKTM combined. Two other contemporary studies tend to support such a figure. Helmut Callis, in an analysis published by the Institute of Pacific Relations, estimated Stanvac's 1936 investment in the Indies at $70 million, and in 1938 Cleona Lewis, in a Brookings Institute study, estimated "American" investment in Netherlands East Indies "oil-producing properties" at $75 million.[36] Considering all of this, $74.4 million appears a reasonable approximation of Stanvac's prewar Indies investment exclusive of petroleum reserves. Any estimate of petroleum reserves is essentially conjecture, and the figure of $150,850,000 has been used solely because that was the estimate given the State Department by Stanvac's board chairman in 1940.[37] At the minimum it gives an indication of the approximate value placed on these reserves by corporate management shortly before the outbreak of war in the Pacific.

[34]Letter, Walden to Hornbeck, September 16, 1940, RG 59, 756D. 94/84.

[35]Letter, Walden to Teagle, April 4, 1940, Exxon files.

[36]Callis, p. 31; Lewis, pp. 230, 588. By all accounts "American" investment in N.E.I. oil was overwhelmingly that of Stanvac in the prewar period. The Standard Oil Company of California had begun exploration in the Indies through the Nederlandsche Pacific Petroleum Maatschappij (NPPM) in 1931, and merged this operation into Caltex (a joint subsidiary formed with the Texas Company in 1936), but no production was underway before the war (Ter Braake, p. 68).

[37]Letter, Walden to Hornbeck, September 16, 1940, RG 59, 756D. 94/84.

Comparative Investment

The simple statement that Stanvac represented the single largest American direct investment in Greater East Asia in 1941 can neither be proven nor disproven with absolute certainty. The possibility always exists that technical discrepancies such as the ones discussed above may have distorted the few comparative studies that do exist. Evidence is good, however, that Stanvac's investments did exceed all other American investments in the region.

Of the available studies, the most comprehensive of the period was that already noted as having been conducted by the Treasury Department in 1943. Reporting was mandatory, book values were used "because no more suitable type of value could be obtained," and property in China, Japan, and Southeast Asia was to be valued as of December 1, 1941.[38] The published results of this survey for East and Southeast Asia along with the original figures submitted by Stanvac are shown in Table A–3. While figures submitted by Stanvac were probably lower than actual value—as discussed above—they were at least prepared under the same ground rules as the others and hence provide a reasonable basis for comparison.[39] A glance at the table shows that Stanvac's reported investment accounted for more than two-thirds of the total for petroleum and exceeded all other categories except manufacturing and agri-

[38]*Census of American-Owned Assets in Foreign Countries [1943]*, p. 59.

[39]Access to the original working papers for this survey, in Record Group 265 at the National Archives Branch in Suitland, is possible only with permission from *each* of the companies whose returns are examined. Permission was obtained from Exxon and Mobil for access to Stanvac's response, but it was not deemed feasible to attempt to gain permission to see *all* of the returns submitted for East and Southeast Asia for the purpose of this study. The total valuation reported to the Treasury for *all* of Stanvac's property in Asia, Africa, and Australia was $126.7 million (Stanvac TFR–500, RG 265).

Table A—3

Value of American Interests in East and Southeast Asia in 1941
(Book value as of December 1, 1941, in millions of dollars)

Country	Stanvac Response	Petroleum	Manufacturing	Mining & Smelting	Public Utility & Transportation	Agriculture	Trade	Finance	Miscellaneous	Total
British Malaya	2.9	#	1.3	3.4	#	13.1	2.7	#	.2	23.9
Burma	.4	#	#	—	—	—	.6	—	—	1.0
China	7.4	13.4	16.8	—	8.0	#	#	#	2.0	40.6
French Asia	.8	#	—	—	—	—	#	—	*	.9
Hong Kong	2.3	2.4	.2	—	-.1	—	.8	-4.5	.5	-.7
Japan	5.5	6.9	18.2	—	*	—	4.1	.8	2.8	32.9
N.E.I.	20.7	34.0	4.6	#	#	35.2	5.4	#	.6	79.8
Philippines	6.0	9.6	16.2	#	30.7	10.4	17.0	8.6	#	94.5
Thailand	.1	#	#	—	—	—	.8	—	#	1.6
Total	46.1	66.3	57.3	3.4	38.6	58.7	31.4	4.9	6.1	274.2 / 266.7

Included in country but not category totals; hence totals do not match.
* Less than $50,000.
SOURCE: U.S. Treasury Department, *Census of American-Owned Assets in Foreign Countries [1943]* p. 71; Stanvac's response is taken from its Form TFR–500 in Record Group 265, National Archives.

culture. Agriculture loomed large in the Indies, but a contemporary study by Callis concluded that Stanvac's oil investment there far exceeded other American interests, with second place held jointly by the rubber plantations of Goodyear and United States Rubber.[40] It appears unlikely that the total East and Southeast Asian investments of any one of these three companies exceeded that of Stanvac. In the case of manufacturing, a 1940 Department of Commerce study reported twenty-seven different American firms engaged in manufacturing in China, eighteen in Japan, and fifteen in the Philippines.[41] Again, it appears unlikely that the investment of any one of these firms exceeded Stanvac's total investment in the region.

Imperfect as this analysis may be, it does provide evidence that the Standard-Vacuum Oil Company did, in fact, represent the single largest American direct investment in East and Southeast Asia in 1941. As noted in the beginning of this Appendix, its facilities were spread throughout the region threatened by Japan, but by far the heaviest investment lay in the oil producing and refining operations of the Netherlands East Indies. In addition, the Indies constituted Stanvac's major

[40]Callis, pp. 31–32. Although the figures differ, this conclusion is compatible with the data contained in the Department of Commerce series on foreign investment; *American Direct Investments—1929*, pp. 26–27; *American Direct Investments—1936*, pp. 16–17; *American Direct Investments—*1940, pp. 16–18. Callis incorrectly included Firestone in his list of American companies with rubber plantations in the Indies; see Allen and Donnithorn, *Western Enterprise in Indonesia and Malaysia*, pp. 119–20, 123.

[41]*American Direct Investments—1940*, p. 16. Remer's 1933 study of investment in China concluded that twenty-eight American firms were engaged in manufacturing, chiefly carpets in the Tientsin-Peiping area (Remer, pp. 285, 289). American interest in Japanese manufacturing was reportedly concentrated in electrical and telephone equipment and the assembly of automobiles (*American Direct Investments—1940, p. 18*). Firms included International Telephone and Telegraph, General Electric, Ford, and General Motors (Wilkins, "Role of U. S. Business," in Borg and Okamoto, eds., *Pearl Harbor as History*, pp. 375–76).

source of supply, providing about half of the total products sold in all of the company's territories prior to the outbreak of war.[42] By any measure, the Indies represented an exceedingly valuable asset to the Standard-Vacuum Oil Company.

[42]Summary of Standard-Vacuum Oil Company 1938–39 Operations, prepared October 2, 1946, Exxon files.

STANVAC'S ASIAN TRADE AND
JAPAN'S OIL SUPPLY

DATA FOR an assessment of Stanvac's position relative to
Japan's oil supply are as limited as those for investment, but
it is possible to reconstruct the broad pattern of trade in the
decade before Pearl Harbor. Statistics clearly confirm the in-
adequacy of Japan's own resources for her growing military
and industrial requirements,[1] and reveal that Japan would not
have needed American oil if she had had access to the full
production of the Indies.[2] This relationship did not escape the
attention of contemporary experts in both the United States
and Japan who freely predicted that Japan would move against
the Indies if cut off from American oil. Furthermore, since
eighty percent of Japan's petroleum imports came from the
United States and nearly fifteen percent from Stanvac and
Royal Dutch-Shell in Southeast Asia,[3] it is clear that Japan's
oil supply could be completely regulated by the American government and these two companies if they chose to act in
unison. This is exactly what occurred in 1940 and 1941.

Aside from these strategic considerations, Stanvac occupied
a position unique among American companies in the Asian oil

[1]Table B-7 clearly shows the excess of consumption over indigenous
production.

[2]Compare total American exports of petroleum products to Japan
in Table B-4 with total Netherlands East Indies exports to all destina-
tions in Table B-3. Although additional N. E. I. oil recorded as destined
for Singapore and other transshipment points probably went to Japan,
it appears that the Japanese share of N. E. I. petroleum exports in the
1930s never ran much more than ten percent.

[3]See Table B-6. American companies exporting to Japan were concentrated on the west coast, and included the Associated Oil Company,
Standard Oil of California, the Union Oil Company, General Petroleum, Richfield, The Texas Company, and Shell.

business. Stanvac owned an extensive distribution network (similar to the organization established by Shell[4]) for the sale throughout Japan and China of refinery products such as gasoline and kerosene.[5] No other American company functioned so extensively inside Japan, and only The Texas Company rivaled Stanvac in China.[6] This made the Standard-Vacuum Oil Company uniquely sensitive to internal measures such as Japan's Petroleum Industry law of 1934.[7] Most other American companies contented themselves with bulk sales to Japanese importers and were relatively unaffected by such actions. Stanvac was also the only American firm whose major source of supply was directly threatened by Japanese ex-

[4]Like the American companies, Royal Dutch-Shell functioned through subsidiaries. Its Far Eastern marketing operations were controlled by the Asiatic Petroleum Company, and its production and refining business in the Indies was handled by the Bataafsche Petroleum Maatschappij (BPM). Asiatic's Japanese marketing was handled by its subsidiary, the Rising Sun Petroleum Company, and three separate subsidiaries functioned in China: the Asiatic Petroleum Company (North China) Ltd., the Asiatic Petroleum Company (South China) Ltd., and the Compagnie Asiatique des Petroles in Yunnan (Allen and Donnithorne, *Western Enterprise in Indonesia and Malaya*, pp. 177–78, and Allen and Donnithorne, *Western Enterprise in China and Japan*, pp. 100, 205–206).

[5]Described in Appendix A.

[6]Stanvac and Rising Sun Petroleum (Shell) were the only two major foreign oil distributors in Japan (Allen and Donnithorne, *Western Enterprise in China and Japan*, p. 206). The Texas Company was well established in China, but clearly ranked behind Stanvac and Asiatic Petroleum in the size of its marketing network (letter, Johnson [Minister to China] to SecState, April 10, 1934, "Report on competition in the China market for petroleum and its products," RG 59, 893. 6363/110).

[7]An estimate of Stanvac sales in Japan is given in Table B-2. While Stanvac controlled roughly a quarter of the Japanese market for gasoline and nearly half the market for kerosene, this represented only a tiny fraction (about three percent) of total Japanese imports of crude and all types of petroleum products. American bulk shipments to Japanese importers accounted for most of the Japanese supply; see Table B-6.

pansion. Standard of California (later Caltex) had begun exploration in the Indies, but in 1941 only Stanvac and Shell were significant producers there.[8] With continuous interference in its distribution business starting in 1934 and a growing threat to its major supply source it is hardly surprising that Stanvac turned to the Department of State and developed a close working relationship with those agencies of the American government concerned with Asian oil. Not only did Stanvac represent the single largest American direct investment in East and Southeast Asia, but it was increasingly threatened by Japan's drive for autarchy and was in a unique position to react by cooperating effectively with the American government in the regulation of the Japanese oil supply. As in the preceding Appendix, what follows is essentially substantiating statistical detail for these general observations.

DATA ON THE broad pattern of prewar oil movement is more readily available than details on Stanvac's portion of the business, but material has been located on both topics and summarized in Tables B–1 through B–7.

Since virtually all the original records of the Standard-Vacuum Oil Company are understood to have been destroyed after it was reorganized out of existence in the early 1960s, data on its prewar Asian trade had to be reconstructed from other sources. A brief analysis of Stanvac's prewar business in the area occupied by Japan was located in the files of the Standard Oil Company (New Jersey) and is condensed in Table B–1.[9] A comparison of these figures with other data strongly suggests that this table includes products purchased from other producers and resold in bulk to importers such as Mitsui in Japan. Direct sales through Stanvac's own distribution network would have been considerably less, but this table does show that Japan was Stanvac's leading market in East and

[8]Ter Braake, p. 68.
[9]Summary of Standard-Vacuum Oil Company pre-World War II "normal" sales volume, prepared October 16, 1947; Exxon files.

TABLE B—1

Distribution of Stanvac Prewar Sales in East and Southeast Asia

(In thousands of 42-gallon barrels; deliveries to Japan and China for 1937, all others 1939)

Country	Gasoline	Kerosene	Diesel Oil	Fuel Oil	Lubricating Oil & Misc.	Total[a]	Percent of Total Sales
Japan[b]	1,491	349	2,359	1,183	208	5,590	43.4
China	527	1,390	630	143	229	2,919	22.6
Philippines	482	258	272	735	115	1,862	14.4
Malaya	152	85	433	348	44	1,062	8.2
N.E.I.	218	405	200	22	153	998	7.7
Indochina	81	137	43	—	41	302	2.3
Thailand	22	38	40	—	17	117	0.9
Burma	—	34	—	—	35	69	0.5
Total	2,973	2,696	3,977	2,431	842	12,919	100.0

[a]Presumably includes stock purchased from other producers and resold in these countries.

[b]Includes Formosa, Korea, and Manchuria.

Source: Summary of Standard-Vacuum Oil Company pre-World War II "normal" sales volume, prepared October 16, 1947; Exxon Files.

Southeast Asia. The fact that Stanvac prewar *sales* were greater in Japan even though prewar *investment* was greater in China is consistent with the conclusions reached by Mira Wilkins in a study of American business and Japanese-American relations.[10]

The details in Table B–2 on Stanvac petroleum sales within Japan derive from two sources. Data for 1931 through 1933 are extracted from a memorandum given to the Department of State by Jersey president Teagle in 1934,[11] and are presumed to be accurate. Data for 1935 through 1939 are taken from an extensive set of statistics on the sources of Japan's oil supply provided to the Secretary of the Treasury by Stanvac board chairman Walden in 1940.[12] These figures are actually Stanvac *shipments* to Japan from the Indies rather than *sales* through the company's distribution outlets in Japan, but it is probable that the two sets of figures are roughly comparable. The data suggest that despite all the furor over Japanese restrictions, Stanvac sales in Japan remained fairly constant until the very end of the decade. Further support for that conclusion is derived from a comparison of Stanvac's total sales in Japan, shown in Tables II–1 and B–1. The reduction at the end of the decade was probably caused by diversion of petroleum to Europe and a shortage of tankers for the Asian trade after the outbreak of war in 1939. At any rate, the statistics do not prove that Stanvac's trade was drastically curtailed by internal Japanese regulation.

For perspective, a summary of petroleum exports from the Netherlands East Indies to Japan from 1931 through 1941 is given in Table B–3. This material was drawn from the U.S.

[10]Mira Wilkins, "The Role of U. S. Business," in Borg and Okamoto, eds., *Pearl Harbor as History*, pp. 362–67.

[11]Memorandum of August 20, 1934, attached to memorandum of conversation, Hornbeck, Phillips, Teagle, and Deterding, August 22, 1934, RG 59, 894.6363/84.

[12]Memorandum of August 6, 1940, Morgenthau Diary, Vol. 292, p. 268. Several pages of these statistics are missing in the Morgenthau Diary, but were located in the copies passed on to the Navy Department; letter, Morgenthau to Knox (Secretary of the Navy), August 13, 1940, file (SC) JJ7-3, CNO Records.

TABLE B—2

STANVAC SALES IN JAPAN, 1931–1939 (EXCLUDING BULK SALES TO OTHER DISTRIBUTORS)

(In thousands of 42-gallon barrels; data for 1934, 1940, and 1941 not available.)

Year	Gasoline	Kerosene	Diesel Oil	Fuel Oil	Lubricating Oil & Misc.	Total
1931[a]	806	n.a.	87	302	179	1,374
1932[a]	905	n.a.	66	331	146	1,448
1933[a]	1,070	n.a.	97	321	106	1,594
1934	n.a.	n.a.	n.a.	n.a.	n.a.	n.a.
1935[b]	810	303	—	782	n.a.	1,895
1936[b]	597	307	5	976	n.a.	1,885
1937[b]	925	239	—	398	n.a.	1,562
1938[b]	1,003	166	—	—	n.a.	1,169
1939[b]	705	281	—	—	n.a.	986

[a]Data for 1931–1933 represent "deliveries into consumption" by Socony-Vacuum in Japan, Korea, and Formosa, excluding deliveries made through Mitsui; derived from memorandum of August 20, 1934, provided to the Department of State by W. C. Teagle; RG 59, 894.6363/84.

[b]Data for 1935–39 represent "importations of . . . petroleum products into Japan" shipped by the Standard-Vacuum Oil Company from the Netherlands East Indies, and presumably do not include stocks that may have been purchased by Stanvac from other shippers and resold in Japan; derived from memorandum of August 6, 1940, provided to the Secretary of Treasury by G. S. Walden; Morgenthau Diary, Vol. 292, p. 268, Roosevelt Library. (See text)

TABLE B—3

NETHERLANDS EAST INDIES EXPORTS OF PETROLEUM AND PRODUCTS TO JAPAN, 1931–1941

(In thousands of 42-gallon barrels)

Year	Crude Oil[a]	Gasoline	Kerosene	Gas Oil & Fuel Oil[b]	Total To Japan	% To Japan	Total Exports[c]
1931	1,074	1	—	56	1,130	4.4	25,692
1932	2,486	376	—	48	2,910	9.9	29,418
1933	—	—	—	2,796	2,796	8.4	33,111
1934	—	—	—	2,718	2,718	7.3	36,982
1935	—	525	—	3,467	3,992	10.3	38,830
1936	—	34	47	3,246	3,327	8.2	40,572
1937	—	260	24	3,507	3,791	8.6	43,955
1938	—	475	99	2,322	2,904	6.4	45,716
1939	—	119	96	2,121	2,336	4.8	48,404
1940	—	242	129	3,118	3,489	7.2	48,256
1941[d]	298	238	195	1,965	2,699	8.3	32,463

[a]Tarakan crude included with "crude" 1931–1932; included with "gas oil & fuel oil" 1933–1941.

[b]Includes diesel oil.

[c]A considerable volume of N. E. I. petroleum exports were recorded as destined for Singapore, Poeloe Bitan, and Poeloe Samboe (islands in the Gulf of Singapore); these were transshipment points, and it is probable that some of these additional products were ultimately shipped to Japan.

[d]January through September only.

SOURCE: U.S. Bureau of Mines, *International Petroleum Trade*: Vol. 2, No. 5, pp. 79–80; Vol. 3, No. 8, pp. 172–74; Vol. 4, No. 9, pp. 179–80; Vol. 5, No. 7, pp. 127–30; Vol. 6, No. 4, pp. 69–72; Vol. 7, No. 3, pp. 70–73; Vol. 8, No. 4, pp. 87–89; Vol. 9, No. 4, p. 147; Vol. 11, No. 1, pp. 12–13; Vol. 11, No. 5, pp. 80–82.

TABLE B—4

UNITED STATES EXPORTS OF PETROLEUM AND PRODUCTS TO JAPAN[a], 1931–1941

(In thousands of 42-gallon barrels)

Year	Crude Petroleum	Gasoline	Kerosene	Gas Oil & Fuel Oil	Lubricating Oil	Total[b]
1931	3,606	1,283	393	5,437	245	10,964
1932	4,877	991	990	4,986	193	12,037
1933	5,533	1,116	516	5,182	180	12,527
1934	6,774	1,224	535	7,930	277	16,740
1935	10,855	905	56	9,335	319	21,470
1936	10,466	1,243	4	9,317	326	21,356
1937	16,668	1,789	471	10,621	475	30,024
1938	22,189	2,085	221	8,327	364	33,186
1939	16,904	2,146	317	9,962	601	29,930
1940	12,373	3,918	114	7,440	1,072	24,917
1941[c]	5,532	2,225	88	3,822	897	12,564

[a]Including Formosa, Korea, Dairen, and Port Arthur.

[b]Excluding small volume of miscellaneous products.

[c]No shipments actually left the U.S. for Japan after the first week of August 1941.

SOURCE: U.S. Department of Commerce, *Foreign Commerce and Navigation of the United States*, volumes for 1931, pp. 727–45; 1934, II, 392–405; 1936, II, 417–30; 1938, pp. 860–75; 1940, pp. 812–26; 1942, pp. 836–50.

TABLE B—5

SOURCES OF PETROLEUM PRODUCTS CONSUMED IN MANCHURIA, 1933
(In thousands of 42-gallon barrels)

Source	Gasoline		Kerosene		Fuel Oil		Lubricants		Total	
	Barrels	Percent	Barrels	Percent	Barrels	Percent	Barrels	Percent	Barrels	Percent
Stanvac	67	24.4	102	27.1	17	19.9	11	18.9	197	24.8
Asiatic Petroleum	90	32.7	93	24.7	67	77.9	3	5.2	253	31.8
Texas Company	38	13.8	40	10.7	1	1.1	3	5.2	82	10.3
Neft (Russian)	33	12.0	58	15.5	*	—	2	3.4	93	11.7
Japanese	45	16.4	83	22.0	*	—	36	62.1	164	20.6
Others	2	.7	—	—	1	1.1	3	5.2	6	.8
Total	275	100.0	376	100.0	86	100.0	58	100.0	795	100.0

*Less than 500 barrels.
SOURCE: Adapted from report of the British Consul General in Mukden to the British Legation in Peking, June 14, 1934, FO371(1934)F5158/142/10, Public Record Office, London.

Sources of Japanese Imports of Petroleum and Products, 1939

(In thousands of 42-gallon barrels)

Shipper and Origin	Crude Oil		Gasoline		Kerosene		Diesel Oil		Fuel Oil		Total	
	Bar-rels	Per-cent	Bar-rels	Per-cent	Bar-rels	Per-cent	Bar-rels	Per-cent	Bar-rels	Per-cent	Bar-rels	Per-cent
Stanvac (N.E.I.)	—	—	705	25.3	281	44.1	—	—	—	—	986	2.9
Asiatic Petroleum (N.E.I., Borneo)	595	3.2	1,012	36.4	138	21.7	1,612	24.5	481	10.0	3,838	11.5
Saghalien Island (Japan)	1,000	5.4	—	—	—	—	—	—	—	—	1,000	3.0
Other Non-U.S.	115	0.6	95	3.4	—	—	—	—	277	5.8	487	1.5
Total Non-U.S.	1,710	9.2	1,812	65.1	419	65.8	1,612	24.5	758	15.8	6,311	18.9

Associated	4,574	24.5	79	2.9	—	—	1,751	26.6	2,182	45.5	8,585	25.6
Standard Oil California	4,884	26.1	57	2.0	77	12.1	—	—	—	—	5,018	15.0
Union	3,078	16.5	121	4.3	109	17.1	465	7.1	92	1.9	3,865	11.5
General Petroleum	865	4.6	215	7.8	2	0.3	1,897	28.8	784	16.4	3,763	11.2
Richfield	1,482	7.9	153	5.5	—	—	—	—	410	8.6	2,045	6.1
Texas	583	3.1	39	1.4	30	4.7	590	8.9	255	5.3	1,497	4.5
Shell	584	3.1	47	1.7	—	—	241	3.7	299	6.2	1,171	3.5
Other U.S.	921	5.0	260	9.3	—	—	29	0.4	15	0.3	1,224	3.7
Total U.S.[a]	16,971	90.8	971	34.9	218	34.2	4,972	75.5	4,037	84.2	27,169	81.1
Total Imports	18,681	100.0	2,783	100.0	637	100.0	6,584	100.0	4,795	100.0	33,450	100.0

[a]There are several discrepancies between these figures for Japanese imports and those for American exports given in Table B-4. Some would presumably be explained by shipments that left one year and arrived the next, but the large discrepancy in gasoline suggests the additional possibility of a different technical classification of certain grades of products in the two countries.

SOURCE: Memorandum of August 6, 1940, provided to the Secretary of the Treasury by G. S. Walden of the Standard-Vacuum Oil Company; Morgenthau Diary, Vol. 292, p. 268, Roosevelt Library, and letter, Morgenthau to Knox, August 13. 1940, file (SC) JJ7-3, CNO Records, Division of Naval History, Navy Yard, Washington, D. C.

Bureau of Mines *International Petroleum Trade* series,[13] probably the most reliable source on this subject available in the United States. Although some additional products were probably transshipped to Japan through Singapore, Poeloe Bintan, and Poeloe Samboe,[14] it is unlikely that Japan received much over ten percent of the output of the Indies in the decade before Pearl Harbor. Presumably because prewar Indies production was dominated by Stanvac and Shell and these firms were primarily interested in distribution of refined products throughout Asia, Africa, and Australia, Japanese bulk importers turned to the United States for supply. That trade, summarized in Table B–4, increased rapidly during the decade, with the added stimulus of Japanese stockpiling. The figures for American exports to Japan are condensed from the Department of Commerce's *Foreign Commerce and Navigation of the United States*,[15] largely because Commerce figures were frequently cited within the State Department, and this represents data as viewed by contemporaries. It is interesting to note that *total* Japanese purchases actually reached a peak in 1938. Thereafter the Japanese bought more gasoline and lubricating oil, but *their purchases of crude fell off*, probably because they lacked both shipping space and foreign exchange.

Table B–6, which details Japanese sources of supply as of 1939, is extracted from the extensive statistics mentioned above, provided to the Secretary of the Treasury by Stanvac

[13]U. S. Department of Interior, Bureau of Mines, *International Petroleum Trade* (Washington, D.C.: Government Printing Office, 1932–; monthly). Early issues of this series are difficult to locate even in governmental records; a complete set was found in the library of the American Petroleum Institute in New York.

[14]Islands in the Gulf of Singapore used for storage and transshipment. A considerable volume of N. E. I. petroleum exports was recorded as destined for these points and cannot be traced further.

[15]U. S. Department of Commerce, Bureau of the Census, *Foreign Commerce and Navigation of the United States* (Washington, D.C.: Government Printing Office, 1802/21–1946; annual, exact title and issuing office vary).

board chairman Walden in 1940.[16] Walden told Secretary Morgenthau that the figures had been obtained from sources "in Japan,"[17] and the extent of detail suggests a reasonable degree of accuracy. Undoubtedly variation occurred during the decade, but the general pattern is consistent with that discussed in a number of contemporary documents. It clearly reveals the heavy dependence of Japan on the United States and the fact that Stanvac and Asiatic Petroleum (Shell) controlled almost all of Japan's supply outside the United States.

For additional perspective, a summary of Japan's oil position from 1931 through September of 1945 is given in Table B–7, condensed from data compiled by the U.S. Strategic Bombing Survey in 1946.[18] The excess of consumption over indigenous production clearly reveals the deficit that so concerned Japanese planners in the 1930s. Several additional points are of interest although not related directly to the subject of this study. Included in the total for indigenous production of refined products is the miniscule output of the synthetic plants in which Japan invested so much time and effort. A peak production of 1.5 million barrels of synthetic oil was reached in 1942, but this was only thirteen percent of what had been planned.[19] Also of interest is the fact that America's *de facto* embargo in July of 1941 had caused total Japanese inventories to decline from 49 million barrels in April to 43 million barrels on December 7.[20] Japanese planners expected

[16]Memorandum of August 6, 1940, Morgenthau Diary, Vol. 292, p. 268; and letter, Morgenthau to Knox, August 13, 1940, file (SC) JJ7-3, CNO Records.

[17]Transcript of meeting, Walden, Morgenthau *et al.*, August 7, 1940, Morgenthau Diary, Vol. 290, p. 52.

[18]U. S. Strategic Bombing Survey, *Oil in Japan's War*, Final Reports 51 and 52 (Statistical Appendix), Report of the Oil and Chemical Division, February, 1946, RG 243. An excellent analysis of this same subject, also based on U. S. Strategic Bombing Survey records, is in Cohen, pp. 133–47. No attempt has been made to reconcile the slight discrepancies between this and the preceding tables since they are not critical to the general conclusions of this Appendix.

[19]Cohen, p. 137.   [20]Ibid., p. 134.

TABLE B—7

JAPANESE OIL POSITION (INNER ZONE), 1931–1945

(In thousands of 42-gallon barrels; Japanese fiscal years April through March)

Fiscal Year	Crude Petroleum		Refined Products[a]			Inventories[b]		
	Imports	Production	Imports	Production	Consumption	Crude	Refined	Total
1931	6,391	1,923	13,303	4,727	14,930	4,919	17,527	22,446
1932	9,136	1,594	14,868	6,298	18,276	3,699	20,526	24,285
1933	10,179	1,419	15,077	7,350	19,422	3,976	23,603	27,579
1934	11,953	1,785	17,181	9,052	22,665	4,040	26,609	30,649
1935	12,829	2,214	20,633	10,130	28,592	3,845	28,919	32,764
1936	15,996	2,458	18,739	10,452	27,699	5,001	31,095	36,096
1937	20,231	2,470	16,651	12,573	29,927	10,467	32,595	43,062
1938	18,404	2,465	14,044	13,142	27,951	12,465	31,891	44,356
1939	18,843	2,332	11,818	11,981	25,261	20,242	31,156	51,398
1940	22,050	2,063	15,110	10,806	28,558	19,901	29,680	49,581
1941	3,130	1,941	5,242	15,997	22,648	20,857	28,036	48,893
1942	8,146	1,690	2,378	16,674	25,794	12,346	25,883	38,229
1943	9,848	1,794	4,652	16,167	27,780	6,839	18,488	25,327
1944	1,641	1,585	3,334	9,615	19,401	2,354	11,462	13,816
1945[c]	0	809	0	1,933	4,582	195	4,751	4,946

[a]Includes aviation gasoline, motor gasoline, diesel fuel, fuel oil, and lubricating oil.
[b]At beginning of each period.   [c]First half of 1945 only (April through September).
SOURCE: U. S. Strategic Bombing Survey, *Oil in Japan's War*, Appendix, pp. 12–15.

this reserve to last two years, by which time they hoped to have free access to oil from the Indies and substantial synthetic production.[21] As noted, synthetics never materialized, but Indies production was quickly restored after the Japanese conquest and reached a peak of 49 million barrels of crude in 1943.[22] Much of this was consumed by the military in Southeast Asia, but a portion went to Japan and is reflected in Table B–7 in the increase of imports shown for 1942 and 1943. The decline thereafter was a result of an increasingly effective American blockade. By the summer of 1945, Japan had almost completely exhausted all of its petroleum reserves.

To return from this digression, the statistics summarized in this Appendix clearly confirm Japan's precarious position with respect to petroleum and the unique position occupied by Stanvac. Not only did the company represent the single largest American direct investment in the region, but it was increasingly threatened by Japan's advance toward the Indies and was uniquely placed to cooperate with the American government in the regulation of Japan's oil supply.

[21]Ibid., p. 135.
[22]Ibid., p. 140.

ESSAY ON SOURCES AND
SELECTED BIBLIOGRAPHY

THIS IS A study from the "outside" rather than the "inside." The logical starting point for an analysis of Stanvac and its relationship with the American government would have been the company's working files for the period prior to World War II, but these are understood to have been destroyed (except for a few legal and financial records) when the company was reorganized out of existence in 1962. Fortunately, the extensive cooperation that developed between Stanvac, Shell, the Department of State, and the British Foreign Office left considerable correspondence in the records of the State Department and Foreign Office, and these proved to be the principal sources for this study. Supplementary material on Stanvac itself was pieced together from a variety of corporate and government materials, and the complex story of the 1941 *de facto* embargo was traced through the records of a number of government agencies and the personal papers of several key participants. Although widely scattered, these sources produced a surprisingly large amount of relevant material, and what follows is a brief description of the archival records that proved most useful. This in turn is followed by a listing of all documentary sources and a selected bibliography of those works that proved most helpful for placing the story in context.

A CONSIDERABLE body of material was located in the National Archives in Washington, D.C., and a useful introduction to those records was found in that institution's *Guide to the Records of the National Archives* (Washington, D.C.: Government Printing Office, 1948). Preliminary inventories have been produced in mimeograph form for most record groups, but the General Records of the Department of State (Record Group 59) are most easily approached from two other directions: a selection of portions of the Department's decimal file for search by use of the subject index available at the Archives, and prior examination of the well indexed *Foreign Relations* series for already published documents in order to identify the decimal files from which they came. On Asian oil problems prior to World War II, the decimal files proved a rich source of incoming and outgoing correspondence and cablegrams, with a multitude of intra-Departmental memoranda attached.

An equally rich source of similar documents was found in the records of the British Foreign Office at the Public Records Office in London. There is an excellent guide to this material in the Public Record Office's *Records of the Foreign Office, 1782–1939* (London: H. M. S. O. 1969), and a detailed index to the British Foreign Office Political Correspondence (Class F.O. 371) is available at the Public Record Office. While almost all relevant material was found in Class F.O. 371, a few sets of records from foreign posts are also available, and complete minutes of Stanvac and Shell negotiations with the Japanese government in 1935 were located in Class F.O. 262, British Embassy and Consular Archives: Japan. These essentially parallel American and British diplomatic records proved especially useful in cross-checking reports for possible discrepancies and differences of viewpoint. They revealed a much closer business-government relationship in Britain than in the United States.

Obtaining information on Stanvac's East Asian operations proved more difficult. Considerable assistance in locating scattered corporate publications came from Dr. Henrietta M. Larson, professor emerita of the Harvard Graduate School of Business Administration and coauthor of the third volume of the *History of Standard Oil Company (New Jersey)*, and Mr. Charles E. Springhorn, retired manager of publications for the Standard Oil Company (New Jersey). Interviews with Mr. Lloyd W. Elliott, managing director of NKPM, Stanvac's production subsidiary in the Indies prior to the war, and Mr. Edward F. Johnson, vice president and counsel for Stanvac in the early 1930s, were especially helpful in providing an understanding of the unique *esprit de corps* of the company's original management team.

Detailed descriptive material on Stanvac came from a variety of sources. With permission of the Exxon Corporation and the Mobil Oil Corporation, a copy of Stanvac's original response to the 1943 Treasury survey of American-owned assets in foreign countries was obtained from Record Group 265, Records of the Foreign Funds Control, at the National Archives; and the controller of Esso Eastern, Inc., was kind enough to permit examination of copies of Esso Standard Eastern Claims for World War II Damages Submitted to the Foreign Claims Settlement Commission of the United States. These two sources provided financial data; other descriptive material was found in Jersey and Socony-Vacuum *Annual Reports*, Jersey's corporate magazine, *The Lamp*, and the *Socony-Vacuum News*, supplemented with material from: Records of the Petroleum Administration for War (Record Group 31), Records of the Bureau of Foreign and Domestic Commerce (Record Group 151), Records of the War Department General Staff, Intelligence Division, Army Attaché Reports (Record Group 165), Records of the Office of Strategic Services, Research and Analysis Branch (Record Group 226), Records of the United States Strategic Bombing Survey (Record Group 243), and a Japanese wartime publication located in the Orientalia Division of the Library of Congress, Tōa kenkyūjo, *Shogaikoku no tai-Shi tōshi* [East Asia Research Institute, *Foreign Investments in China*] (3 vols.; Tokyo, 1942–43), the relevant portions of which were translated by Kook-jin Rhee of Cincinnati. Taken together this material provided a fairly comprehensive picture of Stanvac's prewar operations in East Asia.

For the movement toward an embargo on oil to Japan in 1940 and 1941, the most valuable additional source proved to be the "Diary" of Treasury Secretary Henry J. Morgenthau, Jr., located with his papers at the Franklin D. Roosevelt Library in Hyde Park, New York. The Diary is actually a bound set of volumes of copies of virtually all documents that crossed the Secretary's desk, including verbatim transcripts of meetings and telephone calls and detailed reports on a myriad of subjects in which the Secretary was interested. Since Morgenthau launched a campaign for an oil embargo in mid-1940, the Diary abounds with material on that subject for several months; the quantity declines sharply after Roosevelt told Morgenthau to "get out of the oil business," but there is some material on oil all through 1941. Perhaps most interesting for this study was a long verbatim transcript of a meeting in the fall of 1940 in which Walden briefed Morgenthau on the entire scope of Stanvac's East Asian operations and Japan's dependency on American supplies.

Of special interest in tracing Stanvac's increasing entanglement with the federal bureaucracy in 1940 and 1941 were two sets of files apparently not previously used: One was Records of the Office of the Administrator of the Office of Export Control, found interfiled with the vast and poorly indexed Records of the Foreign Economic Administration (Record Group 169) at the National Archives Branch in Suitland, Maryland. The second was a set of folders marked "Japan—Oil Shipments" from the Foreign Funds Control General Correspondence interfiled with Alien Property Records at the Federal Record Center in Suitland. These required special permission from the Treasury Department to use and proved especially valuable in tracing the entanglement of oil with the freeze of Japanese funds in July 1941. Supplementing all this were the personal papers of a number of government officials active on the oil issue in the 1930s and 1940s. Of the papers examined, the most valuable were those of Joseph C. Grew at the Houghton Library at Harvard University, Stanley E. Hornbeck at the Hoover Institute at Stanford University, and Henry L. Stimson at the Sterling Library at Yale University. The Stimson papers, incidentally, should be used in conjunction with the Records of the Office of the Secretary of War (Record Group 107) for this period, since key files are located in both places.

A considerable body of data was located in Congressional Hearings, publications of the Departments of Commerce and

Interior, and a substantial number of memoirs and auto-
biographies. As a general observation on all this material, it
appears worth noting that even in the absence of a core of
corporate documents it has been possible to piece together a
fairly comprehensive account from other sources. This should
encourage others to undertake similar studies and fill the need
for monographs on the actual relationship between business
and government in the conduct of American foreign relations.

RECORDS AT THE UNITED STATES NATIONAL ARCHIVES,
WASHINGTON, D.C.

International Military Tribunal, Far East, *Record of Proceedings*

Record Group 31, Records of the Petroleum Administration
for War

Record Group 56, General Records of the Department of the
Treasury

Record Group 59, General Records of the Department of State

Record Group 107, Records of the Office of the Secretary of
War

Record Group 151, Records of the Bureau of Foreign and
Domestic Commerce

Record Group 165, Records of the War Department General
Staff; Military Intelligence Division (Army Attaché Reports); and War Plans Division (Joint Board Records)

Record Group 169, Records of the Foreign Economic Administration (Records of the Office of the Administrator of
Export Control)

Record Group 226, Records of the Office of Strategic Services,
Research and Analysis Branch

Record Group 243, Records of the United States Strategic
Bombing Survey

Record Group 265, Records of the Foreign Funds Control
(Stanvac response to 1943 Treasury census of American-
owned assets in foreign countries)

RECORDS AT THE BRITISH PUBLIC RECORD OFFICE, LONDON, ENGLAND

British Embassy and Consular Archives: Japan, Class F. O. 262

British Foreign Office Cabinet Papers, Class F. O. 899

British Foreign Office Political Correspondence, Class F. O. 371

COLLECTIONS OF PERSONAL PAPERS

Norman H. Davis Papers, Library of Congress, Washington, D.C.

W. Cameron Forbes Papers, Houghton Library, Harvard University, Cambridge, Massachusetts.

Joseph C. Grew Papers, Houghton Library, Harvard University, Cambridge, Massachusetts.

Stanley E. Hornbeck Papers, Hoover Institution on War, Revolution and Peace, Stanford University, Stanford, California.

Cordell Hull Papers, Library of Congress, Washington, D.C. D.C.

Nelson T. Johnson Papers, Library of Congress, Washington, D.C.

Frank Knox Papers, Library of Congress, Washington, D.C.

Thomas W. Lamont Papers, Baker Library, Harvard University, Cambridge, Massachusetts.

Jay Pierrepont Moffat Papers, Houghton Library, Harvard University, Cambridge, Massachusetts.

Henry J. Morgenthau, Jr. Papers, Franklin D. Roosevelt Library, Hyde Park, New York.

Robert P. Patterson Papers, Library of Congress, Washington, D.C.

Franklin D. Roosevelt Papers, Franklin D. Roosevelt Library, Hyde Park, New York.

Henry L. Stimson Papers, Sterling Library, Yale University, New Haven, Connecticut.

Miscellaneous Archival Collections

American Petroleum Institute Library, New York, New York (Statistical records).

Archives of the Japanese Ministry of Foreign Affairs, 1868–1945; selected documents microfilmed for the U.S. Library of Congress, 1949–1951, Library of Congress, Washington, D.C.

Esso Standard Eastern, Claims for World War II Damages Submitted to the Foreign Claims Settlement Commission of the United States, copies in the custody of the Controller, Esso Eastern, Inc., New York, New York.

Exxon Corporation Library, New York, New York (Back issues of corporate publications).

Mobil Oil Corporation Library, New York, New York (Back issues of corporate publications).

Record Group 131, Alien Property Records, Federal Records Center, Suitland, Maryland (Foreign Funds Control—General Correspondence, Folders on "Japan-Oil Shipments" only).

Records of the Office of the Chief of Naval Operations, Division of Naval History, Navy Yard, Washington, D.C.

Tōa kenkyūjo, *Shogaikoku no tai-Shi tōshi* (East Asia Research Institute, *Foreign Investments in China*), 3 vols.; Tokyo, 1942–43. (Copy in Japanese Section, Orientalia Division, Library of Congress, Washington, D. C.; relevant portions translated for the author by Kook-jin Rhee).

United States Government Documents

U.S. Congress. *Report of the Joint Committee on the Investigation of the Pearl Harbor Attack,* Senate Document No. 244, 79th Cong., 2d Sess. Washington, D.C.: Government Printing Office, 1946.

U.S. Congress. Joint Committee on the Investigation of the Pearl Harbor Attack. *Pearl Harbor Attack, Hearings,* 79th

Cong., 1st Sess.—79th Cong., 2d Sess. 39 parts; Washington, D.C.: Government Printing Office, 1946.

U.S. Congress. Senate. Special Committee Investigating Petroleum Resources. *American Petroleum Interests in Foreign Countries, Hearings,* 79th Cong., 1st Sess. Part 3 of 6 parts; Washington, D.C.: Government Printing Office, 1946.

U.S. Congress. Temporary National Economic Committee. *Hearings Before the Temporary National Economic Committee.* 84 vols.; Washington, D.C.: Government Printing Office, 1941–45.

U.S. Congress. Temporary National Economic Committee. *Investigation of Concentration of Economic Power, Description of Hearings and Monographs of the Temporary National Economic Committee,* 76th Cong., 3d Sess., Senate Committee Print. Washington, D.C.: Government Printing Office, 1941.

U.S. Congress. Temporary National Economic Committee. *Investigation of Concentration of Economic Power, Temporary National Economic Committee Monograph No. 39; Control of the Petroleum Industry by Major Oil Companies,* by Roy C. Cook, 76th Cong., 3d Sess., Senate Committee Print. Washington, D.C.: Government Printing Office, 1941.

U.S. Congress. Temporary National Economic Committee. *Investigation of Concentration of Economic Power, Temporary National Economic Committee Monograph No. 39-A; Review and Criticism on Behalf of Standard Oil Company (New Jersey) and Sun Oil Company of Monograph No. 39, with Rejoinder by the Monograph Author,* 76th Cong., 3d Sess., Senate Committee Print. Washington, D.C.: Government Printing Office, 1941.

U.S. Department of Commerce. Bureau of Foreign and Domestic Commerce. *American Direct Investments in Foreign Countries—1929,* Trade Information Bulletin No. 731, by Paul D. Dickens. Washington, D.C.: Government Printing Office, 1930.

U.S. Department of Commerce. Bureau of Foreign and Domestic Commerce. *American Direct Investments in Foreign*

Countries—1936, Economic Series No. 1, by Paul D. Dickens. Washington, D.C.: Government Printing Office, 1938.

U.S. Department of Commerce. Bureau of Foreign and Domestic Commerce. *American Direct Investments in Foreign Countries—1940,* Economic Series No. 20, by Robert L. Sammons and Milton Abelson. Washington, D.C.: Government Printing Office, 1942.

U.S. Department of Commerce. Bureau of Foreign and Domestic Commerce. *A New Estimate of American Investments Abroad,* Trade Information Bulletin No. 767, by Paul D. Dickens. Washington, D.C.: Government Printing Office, 1931.

U.S. Department of Commerce. Bureau of the Census. *Foreign Commerce and Navigation of the United States.* Washington, D.C.: Government Printing Office, 1820/21–1946; published annually, but exact title and issuing office vary.

U.S. Department of Interior. Bureau of Mines. *International Petroleum Trade.* Washington, D.C.: Government Printing Office, 1932–; monthly.

U.S. Department of State. *Consular Reports,* "American Petroleum in Foreign Countries," No. 37, January 1884.

U.S. Department of State. *Foreign Relations of the United States.* Washington, D.C.: Government Printing Office, 1852–; exact title varies.

U.S. Department of State. *Foreign Relations of the United States: Japan, 1931–1941.* 2 vols; Washington, D.C.: Government Printing Office, 1943.

U.S. Federal Trade Commission. *The International Petroleum Cartel,* FTC Staff Report Submitted to the Subcommittee on Monopoly of the Select Committee on Small Business, U.S. Senate, 82d Cong., 2d Sess. Washington, D.C.: Government Printing Office, 1952.

U.S. Strategic Bombing Survey. *Oil in Japan's War,* Final Reports No. 51 and 52, Report of the Oil and Chemical Division, February 1946. (Copies in Record Group 243, National Archives, Washington, D.C.)

U.S. Treasury Department. *Census of American-Owned Assets in Foreign Countries* [*1943*]. Washington, D.C.: Government Printing Office, 1947.

Published Collections of Official Documents

Documents on American Foreign Relations, Vols. I–IV. Boston: World Peace Foundation, 1939–1942.

Great Britain. Foreign Office. *British and Foreign State Papers*. London: n.p., 1812/14–.

Ike, Nobutaka, trans. and ed. *Japan's Decision for War: Records of the 1941 Policy Conferences*. Stanford: Stanford University Press, 1967.

League of Nations. *Treaty Series*. 205 vols; Geneva: League of Nations, 1920–1946.

Miller, Hunter, ed. *Treaties and Other International Acts of the United States of America*. 8 vols.; Washington, D.C.: Government Printing Office, 1931–1948.

Royal Institute of International Affairs. *Documents on International Affairs*. New York: Oxford University Press, 1929–; annual.

Diaries and Published Collections of Personal Papers

Cadogan, Sir Alexander. *The Diaries of Sir Alexander Cadogan, 1938–1945,* edited by David Dilks, New York: Putnam, 1972.

Harada, Kumaō. *Saionji-Harada Memoirs*. English translation on reels 49–51 of Microfilms of Documents from the Japanese Ministry of Foreign Affairs, 1868–1945, Library of Congress, Washington, D.C.

Hooker, Nancy H., ed. *The Moffat Papers: Selections from the Diplomatic Journals of Jay Pierrepont Moffat, 1919–1943*. Cambridge: Harvard University Press, 1956.

Ickes, Harold L. *The Secret Diary of Harold L. Ickes*. 3 vols.; New York: Simon and Schuster, 1953–1954.

Israel, Fred L., ed. *The War Diary of Breckinridge Long: Selections From The Years 1939–1944.* Lincoln: University of Nebraska Press, 1966.

Kido, Kōichi. *Kido Kōichi Nikki [Diary].* Abridged English translation on reel WT5 of Microfilms of Documents from the Japanese Ministry of Foreign Affairs, 1868–1945, Library of Congress, Washington, D.C.

Konoye, Fumimaro. *Konoye Memoirs.* English translation in U.S. Congress, Joint Committee on the Investigation of the Pearl Harbor Attack, *Pearl Harbor Attack, Hearings,* Exhibit No. 173, Part 20, pp. 3985-4029. Washington, D.C.: Government Printing Office, 1946.

Nixon, Edgar B. *Franklin D. Roosevelt and Foreign Affairs.* 3 vols.; Cambridge: Belknap Press of Harvard University Press, 1969.

Roosevelt, Elliot, ed. *F. D. R.: His Personal Letters.* 4 vols.; New York: Duell, Sloan and Pearce, 1947–1950.

Roosevelt, Franklin D. *The Public Papers and Addresses of Franklin D. Roosevelt,* compiled and edited by Samuel I. Rosenman. 13 vols.; New York: Random House (Vols. 1–5), Macmillan (Vols. 6–9), Harper (Vols. 10–13), 1938–1950.

CONTEMPORARY ARTICLES AND CORPORATE PUBLICATIONS

Douglas, H. H. "Bit of American History—Successful Embargo Against Japan in 1918 Siberian Occupation Dispute," *Amerasia,* IV (August 1940), 258–60.

Frechtling, Louis E. "Japan's Oil Supplies," *Amerasia,* V (July 1941), 197–201.

———. "Oil and Strategy in the Pacific," *Amerasia,* VI (March 1942), 22–26.

Johnstone, William C. "Export Controls and Far Eastern Policy," *Amerasia,* V (October 1941), 341–43.

Marx, Daniel, Jr. "Shipping Crisis in the Pacific," *Far Eastern Survey,* (May 5, 1941), pp. 87–94.

Schumpeter, E[lizabeth] B. "Problem of Sanctions in the Far East," *Pacific Affairs,* XII (September 1939), 245–62.

Socony-Vacuum Corporation. *Annual Reports.* (Reports for 1930–42)

———. *Socony-Vacuum News* (Issues for 1935–42)

Standard Oil Company (New Jersey). *Annual Reports.* (Reports for 1918–42)

———. *The Lamp.* (Issues for 1918–63)

MEMOIRS AND AUTOBIOGRAPHIES

Acheson, Dean. *Present at the Creation: My Years in the State Department.* New York: Norton, 1969.

Churchill, Winston. *The Second World War.* 6 vols.; Boston: Houghton Mifflin, 1948–53.

Craigie, [Sir] Robert L. *Behind the Japanese Mask.* London: Hutchinson, 1945.

Deterding, Sir Henri, as told to Stanley Naylor. *An International Oil Man.* London: Harper, 1934.

Dirksen, Herbert von. *Moscow, Tokyo, London: Twenty Years of German Foreign Policy.* Norman: University of Oklahoma Press, 1952.

Eden, Sir Anthony. *The Memoirs of Anthony Eden.* 3 vols.; Boston: Houghton Mifflin, 1960–65.

Grew, Joseph C. *Ten Years in Japan: A Contemporary Record Drawn from the Diaries and Private and Official Papers of Joseph C. Grew.* New York: Simon and Schuster, 1944.

———. *Turbulent Era: A Diplomatic Record of Forty Years, 1904–1945,* edited by Walter Johnson. 2 vols.; Boston: Houghton Mifflin, 1952.

Hornbeck, Stanley K. *The United States and the Far East; Certain Fundamentals of Policy.* Boston: World Peace Foundation, 1942.

Hull, Cordell. *The Memoirs of Cordell Hull.* 2 vols.; New York: Macmillan, 1948.

Pratt, Sir John T. *War and Politics in China.* London: Jonathon Cape, 1943.

Stimson, Henry L., and Bundy, McGeorge. *On Active Service in Peace and War*. New York: Harper, 1948.

Tōgō, Shigenori. *The Cause of Japan,* translated and edited by Tōgō Fumihiko and Ben Bruce Blakeney. New York: Simon and Schuster, 1956.

Van Mook, Hubertus J. *The Netherlands Indies and Japan: Battle on Paper, 1940–1941*. New York: Norton, 1944.

Welles, Sumner. *Seven Decisions That Shaped History*. New York: Harper, 1951.

Secondary Works: Books

Allen, George C. *Japanese Industry: Its Recent Development and Present Condition*. New York: Institute of Pacific Relations, 1940.

————. *A Short Economic History of Modern Japan, 1867–1937*. Revised edition; New York: Praeger, 1963.

Allen, George C. and Donnithorne, Audrey G. *Western Enterprise in Far Eastern Economic Development: China and Japan*. London: Allen and Unwin, 1954.

————. *Western Enterprise in Indonesia and Malaya: A Study in Economic Development*. London: Allen and Unwin, 1954.

Aziz, Muhammed A. *Japan's Colonialism and Indonesia*. The Hague: Martinus Nijhoff, 1955.

Barnes, Harry E., ed. *Perpetual War for Perpetual Peace: A Critical Examination of the Foreign Policy of Franklin Delano Roosevelt and Its Aftermath*. Caldwell, Idaho: Caxton Printers, 1954.

Beard, Charles A. *President Roosevelt and the Coming of War, 1941; A Study in Appearances and Realities*. New Haven: Yale University Press, 1948.

Beaton, Kendall. *Enterprise in Oil: A History of Shell in the United States*. New York: Appleton-Century-Crofts, 1957.

Bisson, Thomas A. *American Policy in the Far East, 1931–1941,* revised edition with a supplementary chapter by

Miriam S. Farley. New York: Institute of Pacific Relations, 1941.

Blau, Peter M. *The Dynamics of Bureaucracy.* Chicago: University of Chicago Press, 1955.

Blum, John M. *From the Morgenthau Diaries.* 3 vols.; Boston: Houghton Mifflin, 1959–67.

Borg, Dorothy. *The United States and the Far Eastern Crisis of 1933–1938: From the Manchurian Incident Through the Initial Stage of the Undeclared Sino-Japanese War.* Cambridge: Harvard University Press, 1964.

———. comp. *Historians and American Far Eastern Policy.* New York: East Asia Institute, Columbia University, 1966.

Borg, Dorothy, and Okamoto, Shumpei, eds. *Pearl Harbor as History: Japanese-American Relations, 1931–1941.* New York: Columbia University Press, 1973.

Borton, Hugh. *Japan Since 1931: Its Political and Social Developments.* New York: Institute of Pacific Relations, 1940.

———. *Japan's Modern Century.* New York: Ronald Press, 1955.

Broek, Jan O. M. *Economic Development of the Netherlands Indies.* New York: Institute of Pacific Relations, 1942.

Buckley, Thomas H. *The United States and the Washington Conference: 1921–1922.* Knoxville: University of Tennessee Press, 1970.

Buhite, Russell D. *Nelson T. Johnson and American Policy Toward China, 1925–1941.* East Lansing: Michigan State University Press, 1968.

Burns, James M. *Roosevelt: The Lion and the Fox.* New York: Harcourt, Brace, 1956.

———. *Roosevelt: The Soldier of Freedom.* New York: Harcourt Brace Jovanovich, 1970.

Buss, Claude A. *War and Diplomacy in Eastern Asia.* New York: Macmillan, 1941.

Butler, Sir James R. M. *Grand Strategy, Volume II, September 1939–June 1941,* in *History of the Second World War, United Kingdom Military Series.* London: H. M. S. O., 1957.

245

Butow, Robert J. C. *Tojo and the Coming of the War*. Princeton: Princeton University Press, 1961.

Byas, Hugh. *Government by Assassination*. New York: Knopf, 1942.

Callis, Helmut G. *Foreign Capital in Southeast Asia*. New York: Institute of Pacific Relations, 1942.

Campbell, Charles S., Jr. *Special Business Interests and the Open Door Policy*. New Haven: Yale University Press, 1951.

Carr, Edward H. *International Relations Between the Two World Wars, 1919–1939*. New York: Macmillan, 1947.

Chamberlin, William H. *Japan Over Asia*. Rev. ed.; Boston: Little, Brown, 1939.

Chandler, Alfred D., Jr. *Strategy and Structure: Chapters in the History of American Industrial Enterprise*. Garden City: Doubleday, 1966.

Cheng, Yu-kwei. *Foreign Trade and Industrial Development of China: An Historical and Integrated Analysis Through 1948*. Washington, D.C.: The University Press of Washington, D.C., 1956.

Ch'ien, Tuan-sheng. *The Government and Politics of China*. Cambridge: Harvard University Press, 1950.

Clifford, Nicholas R. *Retreat From China: British Policy in the Far East, 1937–1941*. Seattle: University of Washington Press, 1967.

Cohen, Jerome B. *Japan's Economy in War and Reconstruction*. Minneapolis: University of Minnesota Press, 1949.

Council on Foreign Relations. *The United States in World Affairs, 1934–1935*, by Whitney H. Shepardson. New York: Harper & Brothers, 1935.

Crowley, James B. *Japan's Quest for Autonomy: National Security and Foreign Policy, 1930–1938*. Princeton: Princeton University Press, 1966.

Crozier, Michel. *The Bureaucratic Phenomenon*, translated by the author. Chicago: University of Chicago Press, 1967.

Current, Richard N. *Secretary Stimson: A Study in Statecraft*. New Brunswick: Rutgers University Press, 1954.

Dahl, Robert A. *Modern Political Analysis*. 2nd ed.; Englewood Cliffs: Prentice-Hall, 1970.

de Chazeau, Melvin G., and Kahn, Alfred E. *Integration and Competition in the Petroleum Industry*. New Haven: Yale University Press, 1959.

Dietrich, Ethel B. *Far Eastern Trade of the United States*. New York: Institute of Pacific Relations, 1940.

Divine, Robert A. *The Illusion of Neutrality*. Chicago: University of Chicago Press, 1962.

Drummond, Donald F. *The Passing of American Neutrality, 1937–1941*. Ann Arbor: University of Michigan Press, 1955.

Engler, Robert. *The Politics of Oil: A Study of Private Power and Democratic Institutions*. New York: Macmillan, 1961.

Etzioni, Amitai. *A Comparative Analysis of Complex Organizations*. New York: The Free Press, 1961.

Fairbank, John K. *The United States and China*. 3rd ed.; Cambridge: Harvard University Press, 1971.

Farago, Ladislas. *The Broken Seal: The Story of "Operation Magic" and the Pearl Harbor Disaster*. New York: Random House, 1967.

Feis, Herbert. *The Road to Pearl Harbor: The Coming of the War Between the United States and Japan*. Princeton: Princeton University Press, 1950.

Friedman, Irving S. *British Relations With China, 1931–1939*. New York: Institute of Pacific Relations, 1940.

Furnivall, John S. *Netherlands India: A Study of Plural Economy*. Cambridge: Cambridge University Press, 1939.

Gardner, Lloyd C. *Economic Aspects of New Deal Diplomacy*. Madison: University of Wisconsin Press, 1964.

Gerretson, F. C. *History of the Royal Dutch*. 4 vols.; Leiden: E. J. Brill, 1953–57.

Gibb, George S., and Knowlton, Evelyn H. *The Resurgent Years, 1911–1927*, Volume II in *History of Standard Oil Company (New Jersey)*. New York: Harper & Brothers, 1956.

Gould, James W. *Americans in Sumatra*. The Hague: Martinis

Nijhoff, 1961.

Griswold, A. Whitney. *The Far Eastern Policy of the United States.* New York: Harcourt, Brace, 1938.

Gwyer, J. M. A. *Grand Strategy, Volume III, Part 1, June 1941–August 1942,* in *History of the Second World War, United Kingdom Military Series.* London: H. M. S. O., 1957.

Heinrichs, Waldo H., Jr. *American Ambassador: Joseph C. Grew and the Development of the United States Diplomatic Tradition.* Boston: Little, Brown, 1966.

Hidy, Ralph W., and Hidy, Muriel E. *Pioneering in Big Business, 1882–1911,* Volume I in *History of Standard Oil Company (New Jersey).* New York: Harper & Brothers, 1955.

Hou, Chi-ming. *Foreign Investments and Economic Development in China, 1840–1937.* Cambridge: Harvard University Press, 1965.

Hsü, Immanuel C. Y. *The Rise of Modern China.* New York: Oxford University Press, 1970.

Iriye, Akira. *Across the Pacific: An Inner History of American-East Asian Relations.* New York: Harcourt, Brace & World, 1967.

———. *After Imperialism: The Search for a New Order in the Far East, 1921–1931.* Cambridge: Harvard University Press, 1965.

Johnstone, William C. *The United States and Japan's New Order.* New York: Oxford University Press, 1941.

Jones, Francis C. *Japan's New Order in East Asia: Its Rise and Fall, 1937–1945.* London: Oxford University Press, 1954.

———. *Manchuria Since 1931.* New York: Oxford University Press, 1949.

Kase, Toshikazu. *Journey to the Missouri.* New Haven: Yale University Press, 1950.

Kennan, George F. *American Diplomacy, 1900–1950.* Chicago: University of Chicago Press, 1951.

Kirby, Stanley W. *The Loss of Singapore,* Vol. I in *History of the Second World War: The War Against Japan.* 5 vols.;

London: H. M. S. O., 1957.

Koginos, Manny T. *Panay Incident: Prelude to War*. Lafayette: Purdue University Press, 1967.

Langer, William L., and Gleason, S. Everett. *The Challenge to Isolation, 1937–1940*. New York: Harper & Brothers, 1952.

————. *The Undeclared War, 1940–1941*. New York: Harper & Brothers, 1953.

Larson, Henrietta M., Knowlton, Evelyn H., and Popple, Charles S. *New Horizons, 1927–1950*, Volume III in *History of Standard Oil Company (New Jersey)*. New York: Harper & Row, 1971.

Lewis, Cleona. *America's Stake in International Investments*. Washington, D.C.: Brookings Institution, 1938.

Lockwood, William W. *The Economic Development of Japan: Growth and Structural Change, 1868–1938*. Expanded edition; Princeton: Princeton University Press, 1968.

Louis, William R. *British Strategy in the Far East, 1919–1939*. London: Oxford University Press, 1971.

Lowe, Chuan-hua. *Japan's Economic Offensive in China*. London: Allen and Unwin, 1939.

Lu, David J. *From the Marco Polo Bridge to Pearl Harbor: Japan's Entry Into World War II*. Washington, D.C.: Public Affairs Press, 1961.

Maruyama, Masao. *Thought and Behavior in Modern Japanese Politics*, edited by Ivan Morris. London: Oxford University Press, 1963.

Matloff, Maurice, and Snell, Edwin M. *Strategic Planning for Coalition Warfare, 1941–1942*, in *United States Army in World War II: The War Department* series. Washington, D.C.: Department of the Army, 1953.

Maxon, Yale C. *Control of Japanese Foreign Policy: A Study of Civil-Military Rivalry, 1930–1945*. Berkeley: University of California Press, 1957.

McLean, John G., and Haigh, Robert W. *The Growth of Integrated Oil Companies*. Boston: Graduate School of Business Administration, Harvard University, 1954.

Medlicott, William N. *The Economic Blockade,* in *History of the Second World War: United Kingdom Civil Series.* 2 vols.; London: H. M. S. O., 1952–59.

Mitsubishi Economic Research Bureau. *Japanese Trade and Industry: Present and Future.* London: Macmillan, 1936.

Moody's Industrial Manual. New York: Moody's Investment Service, 1909–; published annually, exact title varies.

Morison, Elting E. *Turmoil and Tradition: A Study in the Life and Times of Henry L. Stimson.* Boston: Houghton Mifflin, 1960.

Morison, Samuel E. *The Rising Sun in the Pacific, 1931–April 1942,* Volume III of *History of U.S. Naval Operations in World War II.* Boston: Little, Brown, 1948.

Morley, James W., ed. *Dilemmas of Growth in Prewar Japan.* Princeton: Princeton University Press, 1971.

Morton, Louis. *Strategy and Command: The First Two Years,* in *United States Army in World War II: The War in the Pacific* series. Washington, D.C.: Department of the Army, 1962.

Nash, Gerald D. *United States Oil Policy, 1890–1964; Business and Government in Twentieth Century America.* Pittsburgh: University of Pittsburgh Press, 1968.

National Foreign Trade Council. *Report of the American Economic Mission to the Far East; American Trade Prospects in the Orient.* New York: National Foreign Trade Council, 1935.

National Planning Association. *Stanvac in Indonesia: Sixth Case Study in an NPA Series on United States Business Performance Abroad.* Washington, D.C.: National Planning Association, 1957.

Nevins, Allan. *Study in Power: John D. Rockefeller, Industrialist and Philanthropist.* 2 vols.; New York: Scribner's, 1953.

Ogata, Sadako N. *Defiance in Manchuria: The Making of Japanese Foreign Policy, 1931–1932.* Berkeley: University of California Press, 1964.

Penrose, Edith T. *The Large International Firm in Develop-*

ing Countries: The International Petroleum Industry. London: Allen and Unwin, 1968.

Perry, Hamilton D. *The Panay Incident: Prelude to Pearl Harbor*. New York: Macmillan, 1969.

Popple, Charles S. *Standard Oil Company (New Jersey) in World War II*. New York: Standard Oil Company (New Jersey), 1952.

Pratt, Julius W. *Cordell Hull, 1933–44*, Volumes XII and XIII of Robert H. Ferrell and Samuel F. Bemis, eds., *The American Secretaries of State and Their Diplomacy*. 2 vols.; New York: Cooper Square Publishers, 1964.

Presseisen, Ernst L. *German and Japan: A Study in Totalitarian Diplomacy, 1933–1941*. The Hague: Martinus Nijhoff, 1958.

Quigley, Harold S. *Far Eastern War, 1937–1941*. Boston: World Peace Foundation, 1942.

Rappaport, Armin. *Henry L. Stimson and Japan, 1931–33*. Chicago: University of Chicago Press, 1963.

Rausch, Basil. *Roosevelt: From Munich to Pearl Harbor*. New York: Creative Age Press, 1950.

Remer, Carl F. *Foreign Investments in China*. New York: Macmillan, 1933.

Romanus, Charles F., and Sunderland, Riley. *Stilwell's Mission to China*, in *United States Army in World War II: China-Burma-India Theater* series. Washington, D.C.: Department of the Army, 1953.

Rōyama, Masamichi. *Foreign Policy of Japan: 1914–1939*. Tokyo: Japanese Council, Institute of Pacific Relations, 1941.

Scalapino, Robert. *Democracy and the Party Movement in Pre-War Japan: The Failure of the First Attempt*. Berkeley: University of California Press, 1953.

Schlesinger, Arthur M., Jr. *The Age of Roosevelt*. 3 vols. to date; Boston: Houghton Mifflin, 1957–60.

Schroeder, Paul W. *The Axis Alliance and Japanese-American Relations, 1941*. Ithaca: Cornell University Press, 1958.

Schumpeter, Elizabeth B., ed. *The Industrilization of Japan and Manchukuo, 1930–1940: Population, Raw Materials and Industry.* New York: Macmillan, 1940.

Sherwood, Robert E. *Roosevelt and Hopkins: An Intimate History.* New York: Harper, 1948.

Shigemitsu, Mamoru. *Japan and Her Destiny: My Struggle for Peace,* translated by Oswald White. New York: Dutton, 1958.

Storry, Richard. *The Double Patriots: A Study of Japanese Nationalism.* London: Chatto and Windus, 1957.

Tansill, Charles C. *Back Door to War: The Roosevelt Foreign Policy, 1933–1941.* Chicago: Henry Regnery, 1952.

Ter Braake, Alex L. *Mining in the Netherlands East Indies.* New York: Institute of Pacific Relations, 1944.

Toland, John. *The Rising Sun: The Decline and Fall of the Japanese Empire, 1936–1945.* New York: Random House, 1970.

Turner, Robert C. *Export Control in the United States during the War and Postwar Periods.* Washington, D.C.: Advanced School of International Studies, 1947.

Vandenbosch, Amry. *The Dutch East Indies: Its Government, Problems and Politics.* 3rd ed.; Berkeley: University of California Press, 1942.

Varg, Paul A. *Open Door Diplomat: The Life of W. W. Rockhill.* Urbana: University of Illinois Press, 1952.

Ware, Edith E. *Business and Politics in the Far East.* New Haven: Yale University Press, 1932.

Weber, Max. *From Max Weber: Essays in Sociology,* translated, edited, and with an introduction by H. H. Gerth and C. Wright Mills. New York: Oxford University Press, 1946.

———. *The Theory of Social and Economic Organization,* translated by A. M. Henderson and Talcott Parsons. New York: The Free Press, 1964.

Wilkins, Mira. *The Emergence of Multinational Enterprise: American Business Abroad from the Colonial Era to 1914.* Cambridge: Harvard University Press, 1970.

———. *The Maturing of Multinational Enterprise: American Business Abroad from 1914 to 1970.* Cambridge: Harvard

University Press, 1974.

Williams, William A. *The Tragedy of American Diplomacy*. Rev. ed.; New York: Dell, 1962.

Williamson, Harold F. *et al. The American Petroleum Industry*. 2 vols.; Evanston: Northwestern University Press, 1959–63.

Wilson, Theodore A. *The First Summit: Roosevelt and Churchill at Placentia Bay 1941*. Boston: Houghton Mifflin, 1969.

Wohlstetter, Roberta. *Pearl Harbor: Warning and Decision*. Stanford: Stanford University Press, 1962.

Woodward, Sir Llewellyn. *British Foreign Policy in the Second World War*, in *History of the Second World War* series. London: H. M. S. O., 1962.

Yanaga, Chitoshi. *Japan Since Perry*. New York: McGraw-Hill, 1949.

Young, Arthur N. *China and the Helping Hand, 1937–1945*. Cambridge: Harvard University Press, 1963.

Young, Marilyn B. *The Rhetoric of Empire: American China Policy, 1895–1901*. Cambridge: Harvard University Press, 1968.

SECONDARY WORKS: ARTICLES, ESSAYS, AND UNPUBLISHED PAPERS

Adams, Frederick C. "The Road to Pearl Harbor: A Reexamination of American Far Eastern Policy, July 1937–December 1938," *Journal of American History*, LVIII (June 1971), 73–92.

Ballantine, Joseph W. "Mukden to Pearl Harbor: The Foreign Policies of Japan," *Foreign Affairs*, XXVII (July 1949), 651–64.

Butow, Robert J. C. "The Hull-Nomura Conversations: A Fundamental Misconception," *American Historical Review*, LXV (July 1960), 822–36.

Crowley, James B. "A New Deal for Japan and Asia: One Road to Pearl Harbor," *Modern East Asia: Essays in In-*

terpretation. Edited by James B. Crowley. New York: Harcourt, Brace & World, 1970.

Dahl, Robert A. "The Concept of Power," *Behavioral Science,* II (July 1957), 201–15.

———. "Power," Volume XII of *International Encyclopedia of the Social Sciences.* Edited by David L. Sills. 17 vols.; New York: Macmillan, 1968.

DeNovo, John A. "The Movement for an Aggressive American Oil Policy Abroad, 1918–1920," *American Historical Review,* LXI (July 1956), 854–76.

———. "Petroleum and the United States Navy Before World War I," *Mississippi Valley Historical Review,* XLI (March 1955), 641–56.

Dixon, D. F. "The Growth of Competition Among the Standard Oil Companies in the United States, 1911–1961," *Business History,* IX (January 1967), 1–29.

Drummond, Donald F. "Cordell Hull, 1933–1944," *An Uncertain Tradition: American Secretaries of State in the Twentieth Century.* Edited by Norman A. Graebner. New York: McGraw-Hill, 1961.

Enos, J. L. "The Mighty Adversaries: Standard Oil Company (New Jersey) and Royal Dutch-Shell," *Explorations in Entrepreneurial History,* X (April 1958), 140–49.

Farnham, Thomas J. "Stanley K. Hornbeck and the Coming of War with Japan, 1937–1941," paper presented at the Duquesne University History Forum, Pittsburgh, Pennsylvania, October 1971.

Herzog, James H. "Influence of the United States Navy in the Embargo of Oil to Japan, 1940–1941," *Pacific Historical Review,* XXXV (August 1966), 317–28.

Hosoya, Chihiro. "Twenty-five Years After Pearl Harbor: A New Look at Japan's Decision for War," *Imperial Japan and Asia: A Reassessment.* Compiled by Grant K. Goodman. New York: East Asian Institute, Columbia University, 1967.

Iriye, Akira. "Japanese Imperialism and Aggression: Reconsiderations," *Journal of Asian Studies,* XXII (August

1963), 469–72; and **XXIII** (November 1963), 103–13.

———. "Japan's Foreign Policies Between World Wars— Sources and Interpretations," *Journal of Asian Studies,* **XXVI** (August 1967), 677–82.

Koistinen, Paul A. C. "The 'Industrial-Military Complex' in Historical Perspective: The Interwar Years," *Journal of American History,* **LVI** (March 1970), 819–39.

Masland, John W. "Commercial Influence Upon American Foreign Eastern Policy, 1937–1941," *Pacific Historical Review,* **XI** (October 1942), 281–99.

Moore, Jamie W. "Economic Interests and American-Japanese Relations," *The Historian,* **XXXV** (August 1973), 551–67.

Morton, Louis. "Germany First: The Basic Concept of Allied Strategy in World War II," *Command Decisions.* Edited by Kent R. Greenfield. Washington, D.C.: Department of the Army, 1960.

———. "Japan's Decision For War," *Command Decisions.* Edited by Kent R. Greenfield. Washington, D.C.: Department of the Army, 1960.

Reed, Peter M. "Standard Oil in Indonesia, 1898–1928," *Business History Review,* **XXXII** (Autumn 1958), 311–37.

Sansom, Sir George. "Japan's Fatal Blunder," *International Affairs,* **XXIV** (October 1948), 543–54.

Williamson, Harold F., and Andreano, Ralph L. "Integration and Competition in the Oil Industry," *Journal of Political Economy,* **LXIX** (August 1961), 381–85.

INDEX

257

Library of Congress Cataloging in Publication Data

Anderson, Irvine H.   1928–
    The Standard-Vacuum Oil Company and United States East Asian
policy, 1933–1941.

    Bibliography: p.
    Includes index.
    1. Standard-Vacuum Oil Company.   2. Corporations, Foreign—
Indonesia.   3. United States—Foreign relations. I.   Title.
HD9569.S83A54   338.7'66'5509598   74–25611
ISBN 0–691–04629–8

Lightning Source UK Ltd.
Milton Keynes UK
UKHW020758150123
415350UK00006B/252